Mathematical Lives

Claudio Bartocci · Renato Betti ·
Angelo Guerraggio · Roberto Lucchetti
Editors

Mathematical Lives

Protagonists of the Twentieth Century
From Hilbert to Wiles

Translated by Kim Williams

 Springer

Editors

Claudio Bartocci
Università di Genova
Dipto. Matematica
Via Dodecaneso 35
16146 Genova
Italy
bartocci@dima.unige.it

Angelo Guerraggio
Università Bocconi
Centro PRISTEM
Via Sraffa 11
20136 Milano
Italy
angelo.guerraggio@unibocconi.it

Renato Betti
Politecnico di Milano
Dipto. Matematica
Piazza Leonardo da Vinci 32
20133 Milano
Italy
renato.betti@polimi.it

Roberto Lucchetti
Politecnico di Milano
Dipto. Matematica
Piazza Leonardo da Vinci 32
20133 Milano
Italy
roberto.lucchetti@polimi.it

Translator:
Kim Williams
Corso Regina Margherita, 72
10153 - Turin (Torino) Italy
http://www.kimwilliamsbooks.com
kwb@kimwilliamsbooks.com

Translation from the Italian language edition: Vite matematiche, edited by C. Bartocci, R. Betti, A. Guerraggio, R. Lucchetti, copyright © 2007 Springer-Verlag Italia, Springer is part of Springer Science + Business Media, All Rights Reserved

The drawings are included with kind permission of Maria Poggi (Queneau, p. 134) and Geraldine D'Alessandris (for all remaining drawings).

Mathematical classification MSC: 01A60, 01A70, 01A80, 03A10, 08-03, 91-03

ISBN 978-3-642-44755-6   ISBN 978-3-642-13606-1 (eBook)
DOI 10.1007/978-3-642-13606-1
Springer Heidelberg Dordrecht London New York

Cover design: deblik

Printed on acid-free paper

Springer is part of Springer Science+Business Media (www.springer.com)

Preface

Mathematical knowledge is increasing at a dizzy rate. In the course of the last 50 years more theorems have been proven that in the preceding thousand years of human history: to give an idea of the order of magnitude, every year in specialized journals alone tens of thousands of research articles are published, and just as many more are made available on the Internet. Even given that the great part of these results are understandable and interesting only to specialists, others represent fundamental intellectual conquests, solving irksome problems or famous conjectures, establishing unexpected connections between various theories or discovering new horizons for research. Furthermore, in many cases these steps forward in mathematics, even those of seemingly limited importance, reverberate in other scientific disciplines, giving rise to innovative conceptual developments or finding surprising technological applications.

Only weak echoes of this fervid intellectual activity reach the general public. The newspapers might carry the news of Andrew Wiles's proof of Fermat's last theorem, or the contorted events surrounding the solution of Poincarè's conjecture by Grisha Perelman, but aside from the sporadic cases mathematics remains more or less ignored. Thus, ironically, in precisely the period of its most florid growth mathematics appears at once extremely fragile, almost a victim of its own excesses of specialisation, relegated to a secondary role in the science of our culture, indeed – in the opinion of the most pessimistic – at risk of extinction as a science in its own right. A few years ago, Gian Carlo Rota commented, "at the end of the second millennium, mathematics seriously risks dying. Among the many threats to its survival, those that loom the largest seem to me to be the crass ignorance of its results, and the widespread hostility towards its practitioners. Both of these are facilitated by the reluctance of mathematicians to push themselves beyond the restricted confines of their own discipline and by their reluctance to translate the esoteric contents into exoteric slogans, which is instead imperative in the age of means of mass communication and public relations".

Whether or not one agrees with these gloomy prophecies, the fact remains that it is not at all easy to coin "exoteric slogans" in order to render the hard-to-digest abstractions of mathematics appetizing to the largest possible number of palates. Physics, biology, and even chemistry can take advantage of concepts that are certain to be attractive – the "secrets of the universe", the "wonders of life", the "mysteries of the molecule" – which, no matter how many times they are served up, still have a grip on the collective imagination (if we can use that expression) and can be used as a point of departure even for works of serious and rigorous popular science. But what are the secrets, the wonders, the mysteries unveiled by mathematics, if not those that appear as such, in all their fascination, only to the eyes of those trained in this discipline?

In an attempt to illustrate the richness of the mathematics of the twentieth century without resorting to slogans or propaganda, the present volume has a new approach: to bring to the forefront some of the protagonists of this extraordinary intellectual adventure, who have put at our disposal new and powerful instruments for investigating the reality around us. There are at least two distinct reasons for making this choice. Above all, the desire to give credit where credit is due. Little has been written on the people – men and women – whose ideas have made possible such deep scientific changes, and they run the risk of remaining in the shadows along with their results. Although many have heard of Russell, Gödel, von Neumann or Nash, how many know about Emmy Nöther, Schwartz, Grothendieck or Atiyah? Secondly, the desire to demonstrate the falsity of a widespread and deeply-rooted belief. It is often held that mathematicians are in every way similar to the extravagant personalities that populate the flying island of Laputa in the Swift's *Gulliver's Travels*. You'll recall that the inhabitants of this land are so lost in mathematical and musical thoughts and concoctions that they can neither talk nor follow anyone else's discussion, and constantly risk falling off some cliff or banging their heads against some obstacle. For this reason they are always accompanied by servants to rescue them, who capture their Masters' attention by touching them on the lips, ears or their eyes with a kind of rattle tied to the end of a stick. Nothing could be further from the truth: mathematicians, bizarre as their behaviour might sometimes appear, have no need at all of solicitous servants to bring them back to reality, because in general their curiosity is vigilant and open to the multiplicities of the world. Many of the portraits contained in this volume present people with strongly charisma, with wide ranging cultural interests, impassioned about defending the importance of their own research, sensitive to beauty, attentive to the social and political problems of their times.

In spite of the inevitable omissions (which we openly knowledge, but as Marcel Schwob observed in the preface to his *Imaginary Lives*, "the art of biography consists precisely in choice"), what we have sought to document is mathematics' central position in the culture – and not only scientific – of our day, in a continuous play of exchanges and references, and correspondences and suggestions. For this reason, in the pages that follow we have made space for not only biographical portraits of the great mathematicians but also for literary texts, which allow us to glimpse this subterranean contiguity. We have even included two intruders (or so they appear, at least at first glance) – Robert Musil and Raymond Queneau –, authors for whom mathematical concepts represented a valuable auxiliary for investigating the modalities of the "new relationship between the phantasmatic lightness of ideas and the weight of the world" (to quote Calvino), to resolve the disagreement between "soul and precision".

Genova, Italy	C. Bartocci
Milano, Italy	R. Betti and A. Guerraggio
April 2010	R. Lucchetti

Editors' Note

The present volume is based on number 50–51 (December 2003–March 2004) of the journal *Lettera matematica PRISTEM*, with modifications, amplifications and significant additions.

We are grateful to Maria Poggi and Geraldine d'Alessandris for their nice and expressive drawings.

Contents

Hilbert's Problems

A Research Program for "Future Generations"

Umberto Bottazzini

In an isosceles triangle, if the ratio between the base angle and the vertex angle is an algebraic but irrational number, is the ratio between the base and the side always transcendent?

The simplicity of this question is deceiving. This is not an exercise in Euclidean geometry that can be solved by a bright student, but is instead a translation into geometric terms of the fact that the exponential function $\exp(i\pi z)$ must always be a transcendental number for irrational algebraic values of z. David Hilbert thought that this was "highly probable", although providing a proof of it seemed to be an "extremely difficult" undertaking. Thus he added it to the list of problems for "future generations" that he presented in Paris on 8 August 1900 during the second International Congress of Mathematicians.

"Who of us would not be glad to lift the veil behind which the future lies hidden; to cast a glance at the next advances of our science and at the secrets of its development during future centuries?", exclaimed Hilbert at the beginning of his talk. "What particular goals will there be toward which the leading mathematical spirits of coming generations will strive? What new methods and new facts in the wide and rich field of mathematical thought will the new centuries disclose?"

It was a unique moment. The Congress, on the cusp of two centuries, offered the mathematician from Göttingen a chance to "look over the problems which the science of today sets", and invite the mathematicians of "future generations" to put themselves to the test. His talk defined an epoch. However, for those who imagine Hilbert reading his talk, soon to be become legend, to a hall filled with the most authoritative mathematicians of the times, reports about the congress contain some surprises. According to Gino Fano, the audience was not actually very large. Many of the participants didn't attend. There were about ten Italians: Peano and his followers (Amodeo, Padoa, Vailati), a couple of high school teachers, and then Levi-Civita and Volterra, who gave the opening address. Of the Germans, neither Klein nor Nöther were present, nor were any of the mathematicians from Berlin. Even among the French, leading mathematicians such as Hermite, Picard, Jordan, Goursat, Humbert and Appell failed to attend the Congress sessions. Hilbert's talk was one of those in the section entitled *Bibliographie et Histoire. Enseignement et methodes* with historian Moritz Cantor presiding. Hilbert limited himself to presenting about 10 of the 23 problems that appeared in the text prepared for publication. Thus, the volume of the acts of the Congress contains a note saying that "an enlarged version of the talk of Mr Hilbert, because of its great importance, has been included among the lectures".

David Hilbert

Hilbert's observations regarding methodology in the introduction to the problems shed light on his conception of mathematics and its development. "The deep significance of certain problems for the advance of mathematical science in general and the important role which they play in the work of the individual investigator are not to be denied", he said, and continued, "An old French mathematician said: 'A mathematical theory is not to be considered complete until you have made it so clear that you can explain it to the first man whom you meet on the street'. This clearness and ease of comprehension, here insisted on for a mathematical theory, I should still more demand for a mathematical problem . . . [it] should be difficult in order to entice us, yet not completely inaccessible, lest it mock at our efforts". The failure to solve a problem often depends on "our failure to recognize the more general standpoint from which the problem before us appears only as a single link in a chain of related problems". Once the right level of generality is found, not only does the problem show itself to be more accessible, but often the right methods to solve problems related to it also appear. An unlimited faith in the capacity of human reason led Hilbert to formulate a kind of "general law" for our thinking, to establish a kind of axiom that it was possible to find a solution for any mathematical problem whatsoever. "In mathematics there is no *ignorabimus*", he stated optimistically – perhaps *too* optimistically – in defiance the notorious statements by Emil Du Bois-Reymond.

Among the classic problems, Hilbert noted Johann Bernoulli's brachistochrone problem, which had led to the birth of the calculus of variations, and Fermat's last theorem, which gave rise to Kummer's theory of ideal numbers and their generalisations to all algebraic fields through the work of Dedekind and Kronecker. The three body problem, which in recent times had led Poincaré to the discovery of "fruitful methods and far-reaching principles", was of an entirely different nature. For Hilbert, as mathematical problems Fermat's theorem and the three body problem were situated at "opposite poles": "the former a free invention of pure reason, belonging to the region of abstract number theory, the latter forced upon us by astronomy and necessary to an understanding of the simplest fundamental phenomena of nature". Like the three body problem, Hilbert observed: "Surely the first and oldest problems in every branch of mathematics spring from experience and are suggested by the world of external phenomena". This was the case for the operations of counting or the classic problems of geometry, the duplication of the cube or the quadrature of the circle. However, "in the further development of a branch of mathematics, the human mind, encouraged by the success of its solutions, becomes conscious of its independence. It evolves from itself alone, often without appreciable influence from without, by means of logical combination, generalization, specialization, by separating and collecting ideas in fortunate ways, new and fruitful problems, and appears then itself as the real questioner". Thus Hilbert explained the origins of the problem of the distribution of prime numbers, Galois's theory of algebraic invariants, and the theories of Abelian and automorphic functions. In short, "almost all the nicer questions of modern arithmetic and function theory".

He goes on, "In the meantime, while the creative power of pure reason is at work, the outer world again comes into play, forces upon us new questions from actual experience, opens up new branches of mathematics, and while we seek to conquer these new fields of knowledge for the realm of pure thought, we often find the answers to old unsolved problems and thus at the same time advance most successfully the old theories. And it seems to me that the numerous and surprising analogies and that apparently prearranged harmony which the mathematician so often perceives in the questions, methods and ideas of the various branches of his science, have their origin in this ever-recurring interplay between thought and experience".

Thus, in the continual interaction between unfettered creations of the mind and knowledge of the phenomena of the external world, Hilbert finds the fundamental dynamic of mathematical development, along with the driving force behind the process of the mathematisation of the other sciences. The rigour of the proofs, a particular characteristic of mathematics – considered by Hilbert to be "a universal philosophical necessity of our understanding" – was also required in the treatment of the most delicate problems of analysis and those questions that originate in the external world, in the world of empirical experience.

The text of Hilbert's Paris talk belies the caricature often used to portray the Hilbertian concept of mathematics, reducing it to a purely formal game with meaningless symbols. To be sure, "to new concepts correspond, necessarily, new signs", observed Hilbert. But these signs are chosen so as to recall the phenomena that generated them. Thus, for example, "the arithmetical symbols are written diagrams and the geometrical figures are graphic formulas; and no mathematician could spare these graphic formulas". On the other hand, he continued, "we do not habitually follow the chain of reasoning back to the axioms in arithmetical, any more than in geometrical discussions". When addressing a new problem, "we apply, especially in first attacking a problem, a rapid, unconscious, not absolutely sure combination, trusting to a certain arithmetical feeling for the behaviour of the arithmetical symbols, which we could dispense with as little in arithmetic as with the geometrical imagination in geometry". He had put this vision to the test in his own research on the foundations of geometry, the subject of a course he taught and of the volume entitled *Grundlagen der Geometrie*, which appeared in 1899 as a *Festschrift* on the occasion of the inauguration of the monument to Gauss and Weber in Göttingen.

In the introductory explanation to the *Grundlagen*, Hilbert declared that he considered "three different systems of objects", called respectively points, lines and planes. He added that "the exact and complete description" of the relations between the three were entrusted to the axioms. The logician Gottlob Frege objected that in so doing, the axioms were given the task usually assigned to the definitions. Frege was convinced that the axioms of geometry were true statements, the knowledge of which "grows out of a cognitive source that is of an extra-logical nature, which we might call spatial intuition".

For Hilbert, in contrast, the axioms were not statements that were true in themselves. The criteria for establishing the truth and existence of mathematical objects was entrusted to the proof of the non-contradictoriness of the axioms (and of their consequences). He retorted to Frege's criticism, saying, "All theories are only a frame, a layout of concepts that are together with their necessary mutual relationships", which can be applied to "infinite systems of fundamental entities". These fundamental entities can be thought of arbitrarily. In order to obtain all of the propositions of the theory it was sufficient that the relationships between the fundamental entities be established by the axioms. The axiomatic method shed light on the deductive weave, the way in which axioms and theories depend on each other. In Hilbert's eyes, this was its essential value. Of course, if we want to apply a theory to the world of phenomena, then "a certain amount of good intention and certain sense of measure" was necessary. Instead, applying an axiomatic theory to phenomena other than those for which the theory was ideated required "an enormous amount of bad intention".

The problems proposed by Hilbert touched on a variety of questions: in the first place, the foundations of analysis (problems 1 and 2), geometry (problems 3, 4 and 5), and the axiomatisation of physical theories (problem 6). The first problem concerned the nature of continuum: "every infinite system of real numbers, that is, every infinite set of numbers (or points) is either equivalent to the set of all natural numbers 1, 2, 3, ... or equivalent to the set of all real numbers, and as a consequence, of the continuum". From the proof would have followed the proof of Cantor's *continuum hypothesis*, according to which 'the number of real numbers is the level of infinity immediately above countable infinity'. According to Hilbert, the "key to the proof" might perhaps lie in Cantor's statement that every infinite set could be well ordered. The set of real numbers, in natural order, was certainly not a well-ordered set. However, Hilbert asked, was it possible to find for that set a different order such that each of its subsets had a prime element? In other words, was it possible to find a well-ordered sequence for the continuum?

Before any mathematician could respond to that question, Bertrand Russell pointed out an antinomy that posed a serious threat to the foundations of the entire construction of Cantor's set theory. The question posed by Hilbert thus came to be entwined with the more general question of the basic principles of Cantor's set theory and gave rise to an enormous mass of studies both logical and foundational, in which many of Hilbert's students and collaborators were involved, beginning with Ernst Zermelo who, in 1904, provided a first axiomatisation of set theory and shed light on the role of the so-called "axiom of choice". With particular regard to Cantor's continuum hypothesis, a first significant result was obtained by Kurt Gödel who, in 1938, proved that the (generalised) continuum hypothesis could not be disproved from the axiom of choice and other axioms of set theory. However, it was not until 1963 that Paul Cohen demonstrated that it couldn't be proven by those axioms either.

The second problem proposed by Hilbert was intimately related to the first. In the *Grundlagen* he had shown that the non-contradictoriness of the axioms of Euclidean geometry was related to the axioms of the arithmetic of real numbers, in the sense that, as he explained, "every contradiction in the deductions of the axioms of geometry must be traced back to arithmetic" of real numbers. Thus, he continued, "this makes a direct method for the proof of non-contradictoriness of the axioms of arithmetic necessary", essentially the axioms for the usual rules of calculation with the addition of an axiom of continuity (that is, the *axiom of Archimedes* and a new *completeness axiom* stated by Hilbert in a then recent work which established the impossibility of an Archimedean extension of the line of real numbers and modified an essential point of the system of axioms established in the first edition of the *Grundlagen*).

Hilbert attributed a decisive role to the proof of non-contradictoriness as a criterion for the existence of mathematical objects. A few months earlier, in reply to Frege's criticism of the axiomatic formulation of the *Grundlagen*, he had written, "If arbitrarily established axioms are not contradictory in any of their consequences, then they are true, and then defined entities exist by means of those axioms. I consider this to be the criterion for truth and existence". He now declared publicly, "If contradictory attributes are assigned to a concept, I say that mathematically that concept does not exist". Hilbert had amazed the world with proofs of an existential nature some 10 years earlier (the 1888 *basis theorem* and the 1890 *theorem of zeroes*). The hoped-for proof of the non-contradictoriness of the axioms of arithmetic would have proven the existence of both the real numbers and the continuum. Because the consistency of geometry and of analysis could be traced back to that of arithmetic, the direct proof of the non-contradictoriness of the axioms of arithmetic would have guaranteed the consistency of the whole of mathematics. The second problem was in fact a statement of this ambitious program, which Hilbert and his students would pursue through the 1920s, before Gödel's incompleteness theorem of 1931 proved that the task was an impossible one in terms of how it had been formulated by Hilbert, which led to its being drastically revised.

The next three problems were inspired by Hilbert's own research on the foundations of geometry. In the *Grundlagen* Hilbert had shown that in plane geometry the axioms of congruence (without resorting to the axiom of continuity) were sufficient to prove the congruence of straight line figures. Gauss had already noted that, instead, the proof of theorems of solid geometry such as that of Euclid – prisms of equal height and triangular bases are proportional to their bases – depends on the method of exhaustion, that is, in the final analysis, to an axiom of continuity. In problem three, Hilbert asked to be shown "two tetrahedra of equal bases and equal heights that cannot be subdivided into congruent tetrahedra". The proof was produced 2 years later by Hilbert's student Max Dehn (1878–1952).

Another of Hilbert's students, Georg Hamel (1877–1954) had successfully taken on the fourth problem. Hilbert had drawn attention to the geometry developed by Minkowski in the 1896 *Geometrie der Zahlen*, in which all of the axioms of ordinary geometry were valid (including the axiom of parallels) with the exception of the axiom of congruence of triangles, which was replaced by the axiom of triangular inequality. Hilbert himself, in 1895, had studied a geometry in which all of the axioms of Minkowski's geometry were valid, except the axiom of parallels. Convinced of their importance for number theory, theory of surfaces and the calculus of variations, Hilbert now called for a systematic study of the geometries in which all of the axioms of Euclidean geometry were valid except for the axiom of triangular congruence (axiom III, 5 of *Grundlagen*), which was substituted by triangular inequality, taken as a particular axiom. Hamel proved that the only possible geometries were elliptic (in the case of an integer plane) or hyperbolic such as the kind studied by Minkowski and Hilbert. The problem was in any case formulated by Hilbert in terms that were quite vague, and in the decades that followed this gave rise to numerous studies on particular classes of geometries.

In his work on continuous transformation groups, Lie had established a system of axioms for geometry and resolved the problem of how to determine all the *n*-dimensional manifolds that admit a group of rigid motions, in other words, the problem posed by Riemann and Helmholtz as to the characterisation of the rigid motions of bodies. Lie had assumed that the transformations of his groups would be differentiable functions. In 1898 Klein had expressed doubts as to whether this hypothesis was necessary, and now, in problem five, Hilbert took up the question once more, asking himself if, as far as the axioms of geometry were concerned, the hypothesis of differentiability was inevitable or if instead this was a consequence of other geometric axioms.

More than an actual problem, Hilbert's sixth problem provided a guideline for research. Using as a model the studies on the principles of arithmetic and geometry, Hilbert invited mathematicians "to treat in the same manner, by means of axioms, those physical sciences in which mathematics plays an important part". What he had in mind was the kind of probabilistic concepts introduced by Clausius and Boltzmann in the kinetic theory of gas, and Mach's and Boltzmann's research on the foundations of mechanics. Starting at the beginning of the twentieth century, first with Minkowski and then, after his friend's premature death in 1909, with his assistants, Hilbert studied problems of theoretical physics with growing interest for a couple of decades. He taught courses and gave lectures on particular topics; he published important works, such as his 1915 paper, which appeared just a few weeks after that of Einstein, in which he obtained the equations for general relativity and exhorted his students and collaborators to engage in this kind of research. Also situated in that field was Emmy Nöther's 1918 theorem – regarding the calculus of variations – of fundamental importance in modern mathematical physics, which put the number of parameters of a subgroup of invariants for lagrangian systems in relation to number of laws of conservation that could be derived for those systems. With Richard Courant he wrote the treatise *Matematische Methoden der Physik* (1924), which became a classic. Books by his students, such as *Gruppentheorie und Quantenmechanik* (1928) by Hermann Weyl, and *Mathematische Gundlagen der Quantenmechanik* (1932) by John von Neumann are considered to be among the most significant results produced in the spirit of Hilbert's sixth problem. However, as Weyl himself admitted, in spite of the great results achieved, "Hilbert's plans in physics never matured".

As far as probability theory is concerned, the axiomatisation hoped for by Hilbert took shape in the Russian school of Bernstein and Kolmogorov, in the context of modern measure theory.

After the problems concerning foundations, Hilbert passed to a consideration of specific problems, beginning with number theory, the discipline around which his research in recent decades had focussed, culminating in the publication of the *Zahlbericht* (1897). This is reflected above all in problem seven, which we mentioned earlier and other problems correlated to it (which were solved in the 1930s by Gelfond), and in problem eight regarding the distribution of the prime numbers and the *Riemann hypothesis*, which is perhaps the most important conjecture still open in mathematics today.

These two problems are among those that Hilbert presented during his talk in Paris. The others he mentioned were the first, second, sixth, thirteenth, sixteenth, nineteenth and twenty-second. With the complex problems comprised between the ninth and the eighteenth, he passed from number theory to problems of algebra or algebraic geometry. The tenth problem, for example, asked for a procedure that would be able to determine by means of a finite number of operations whether or not a given Diophantine equation with n unknowns has integer solutions. Instead, the 13th asked for a proof showing that the generalised seventh-degree equation can be solved with functions of only two parameters. Establishing rigorous foundations for Schubert's enumerative calculus in geometry was the aim of the 15th problem, while problem 16 regarded the topology of algebraic curves and surfaces. The next two problems were also geometric in nature. Extending a theorem established in the final chapter of the *Grundlagen*, which was dedicated to the possibility of constructions with compass and straightedge, problem 17 asked if for any definite form (that is, a rational integer function of n variables that takes only non-negative values over the reals) could be expressed as a quotient of the sum of squares. Problem 18 asked for an extension of Poincaré's (and Klein's) results regarding Lobachevsky's plane (and space) groups of motions to n-dimensional Euclidean space. To this was correlated a question that was "important to number theory and perhaps sometimes useful to physics and chemistry: How can one arrange most densely in space an infinite number of equal solids of given form, e.g., spheres with given radii or regular tetrahedra with given edges. . .?"

In the final group of problems, Hilbert took topics in analysis into consideration. In the 19th problem, he questioned "whether all solutions of regular variation problems must necessarily be analytic functions", while the 20th problem regarded whether or not there exist solutions to partial differential equations with certain boundary conditions.

In the 21st problem, inspired by Riemann's and Fuchs's results, Hilbert asked for proof of the existence of a linear differential equation with given singular points and monodromic group. The next to last problem regards an extension of Poincaré's uniformisation in the theory of automorphic functions. Last but not least, with the 23rd problem Hilbert calls for "further development of the calculus of variations".

Looking at the 23 problems as a whole, it is possible to see that the original studies outline the scheme of development for some of the most important branches of twentieth-century mathematics. While some of the problems were stated clearly and precisely, in other cases Hilbert instead urged young mathematicians to create new theories or research programs. From this point of view, the more than 60 doctoral theses written under his direction between 1898 and 1915 are revealing. Eleven of his students wrote a thesis on questions of number theory, and three of their topics were related to the 12th problem – Hilbert was inspired by Kronecker's *Jugendtraum*, or youthful dream – which regarded the development of the parallels between fields of algebraic numbers and fields of algebraic functions. Ten of the theses dealt with the foundations of geometry and problems of algebraic geometry in strict correlation to the16th problem. However, almost half of his doctoral students deal with the topics in analysis that Hilbert was predominantly interested in up to the First World War, above all the calculus of variations (in particular with *Dirichlet's principle*) and the theory of integral equations. In the 1920s, five of the nine theses overseen by Hilbert dealt with the foundations of mathematics and proof theory. These concerned the development of ideas outlined in the second problem, to which Hilbert dedicated the final phase of his work, tying his name to the so-called formalist program of the foundations of mathematics.

Hilbert's 23 Problems

At the second International Congress of Mathematicians, which took place in Paris in 1900, David Hilbert presented 23 problems that were unsolved at the time in various areas of mathematics. In his opinion, these were the problems to which the attention of the researchers of the new century would be drawn.

1. *Cantor's problem of the cardinal number of the continuum (the continuum hypothesis)*: Is there set whose size is strictly between that of the integers and that of the continuum? In 1938 Gödel proved that the continuum hypothesis is consistent with Zermelo–Fraenkel set theory; in 1963 Cohen proved that its negation is as well.

2. *The compatibility of the arithmetical axioms*: Gödel showed in 1931 that no proof of its consistency can be carried out within a system as rich as arithmetic.

3. *The equality of two volumes of two tetrahedra of equal bases and equal altitudes*: Max Dehn found a counterexample in 1902.

4. *Problem of the straight line as the shortest distance between two points*: Construct all the metric geometries in which the lines are geodesics. Solved in 1901 by Georg Hamel.

5. *Lie's concept of a continuous group of transformations without the assumption of the differentiability of the functions defining the group*: Is it possible to avoid the hypothesis that the transformations are differentiable to introduce the concept of continuous transformation groups according to Lie? Solved for particular transformation groups by John von Neumann in 1933 and, in the general case, by Andrew Gleason and independently by Deane Montgomery and Leo Zippin in 1952.

6. *Mathematical treatment of the axioms of physics*: In particular, the axiomatisation of those areas, such as mechanics and probability theory, in which mathematics is essential. Results were produced by Caratheodory (1909) in thermodynamics; von Mises (1919) and Kolmogorov (1933) in probability theory; John von Neumann (1930) in quantum theory; Georg Hamel (1927) in mechanics.

7. *Irrationality and transcendence of certain numbers*: In particular, if a^b is transcendent when base a is algebraic and exponent b is irrational. An affirmative answer was given by Gelfond in 1934 and (independently) by Schneider in 1935.

8. *Problems of prime numbers*: In particular Riemann's hypothesis on the zeroes of Riemann's "zeta function" relative to the distribution of primes.

9. *Proof of the most general law of reciprocity in any number field*: Resolved for a special case by Teiji Takagi in 1920, and more generally by Emil Artin in 1927.

10. *Determination of the solvability of a Diophantine equation*: Is there a univeral algorithm for their solution? A negative answer was provided by Yuri Matiyasevich in 1970.

11. *Quadratic forms with any algebraic numerical coefficients*: Solved by Helmut Hasse in 1923.

12. *Extension of Kroneker's theorem on abelian fields to any algebraic realm of rationality*: Solved by Shimura and Taniyama in 1959.

13. *Impossibility of the solution of the general equation of the seventh degree by means of functions of only two arguments*: Generalises the impossibility of solving a fifth-degree equation by roots. Answered in the negative by Kolmogorov and Arnol'd in 1961: a solution is possible.

14. *Proof of the finiteness of certain complete systems of functions*: A first counter-example was provided by Nagata in 1958.

15. *Rigorous foundation of Schubert's enumerative calculus*: precisely determine the limits of the validity of the numbers that Hermann Schubert had determined on the basis of the principle of special position, by means of his enumerative calculus. Solved.

16. *Problem of the topology of algebraic curves and surfaces*: In particular, developing Harnack's methods and Poincaré's theory of limited cycles.

17. *Expression of definite forms by squares*: In 1927 Emil Artin proved that a positive definite rational function is the sum of squares.

18. *Building up of space from congruent polyhedra*: Solved (but Penrose found non-periodic solutions).

19. *Are the solutions of regular problems in the calculus of variations always necessarily analytic?* Partially solved in 1902 by G. Lötkeyeyer and more generally in 1904 by S. Bernstein. General solution by De Giorgi in 1955 and by J.F. Nash Jr. independently some months later.

20. *The general problem of boundary values*: do variational problems with particular boundary conditions have solutions? Resolved.

21. *Proof of the existence of linear differential equations having a prescribed monodromic group*: Partially resolved by Hilbert in 1905, and by Deligne for other special cases in 1970. A negative solution was found by Andrej Bolibruch in 1989.

22. *Uniformization of analytic relations by means of automorphic functions*: Solved in 1907 by Paul Koebe.

23. *Further development of the methods of the calculus of variations.*

The Way We Were

The Protagonists of the "Italian Spring" in the First Decades of the Twentieth Century

Giorgio Bolondi, Angelo Guerraggio, and Pietro Nastasi

After the trial run in Zürich (1897), the *International congresses of mathematicians* officially began with Paris (1900) and Heidelberg (1904). The third was held in Rome (1908). This order was not random, nor was it dictated only by contingencies. The fact is, at the beginning of the twentieth century Italian mathematics was considered the third world "power", immediately after the great and traditional French and German schools. The same classification holds, almost completely unchanged, at the beginning of the 1920s. American mathematician G. D. Birkhoff, particularly attentive to the situations of European research centres (and interested in consolidating collaborations with them for the definitive launch of the mathematics of the United States) does not hesitate to place Rome immediately after Paris, even before Göttingen.

But who was in Rome in those years? Who were the mathematicians who made it possible for Italian mathematics to compete with the more famous schools of Europe (and therefore, for the moment, of the world)?

After the Italian Unification and the successive transfer of the capital to Rome, the political leaders had made it part of their policy to bring the most vivacious aspects of culture to the city. This also included the scientific culture. The first to arrive among the mathematicians referred to by Birkhoff was Guido Castelnuovo, who transferred to Rome in 1891. In truth a little more than 30 years would have to pass before the Italian school of algebraic geometry would regroup in the capital, but in the end – though with a bit more effort and ill will that foreseen – it made it. In 1923, Federigo Enriques and Francesco Severi came to Rome as well. Vito Volterra arrived to the capital from Turin in 1900 and was immediately charged with giving the inaugural address at the beginning of the academic year. The choice of argument was not a given: Volterra chose to speak *"On the attempts to apply mathematics to the biological and social sciences"*. Tullio Levi-Civita would arrive in Rome shortly after Volterra, in 1909, but for the time being he didn't want to leave the tranquillity of Padua. He transferred only after World War I (1918), after he had married, and after a first period spent in Rome, following the defeat at Caporetto.

C. Bartocci et al. (eds.), *Mathematical Lives*,
DOI 10.1007/978-3-642-13606-1_2, © Springer-Verlag Berlin Heidelberg 2011

Vito Volterra

Great mathematicians all – Castelnuovo, Enriques, Severi, Volterra, Levi-Civita – but as we shall see, with something extra.

The most spectacular phase of the development of the Italian school of algebraic geometry is identified with Castelnuovo, Enriques and Severi in the last decade of the nineteenth century and the first two of the twentieth. They are the ones who ensured for Italian mathematics the pre-eminence that A. Brill publicly recognized in the preface to Severi's *Vorlesungen über algebraische Geometrie* (1921), inviting young German scholars to take note in order to accept the challenge and return to the top of research. Nor was other international recognition lacking. The Bordin Prize of the Paris Academy of Sciences was awarded in 1907 to Enriques and Severi (and in 1909 to G. Bagnera and M. de Franchis), for research in what was to become "Italian geometry". In 1908, a commission made up of M. Nöther, E. Picard and C. Segre had awarded the Guccia Medal to Severi. The publication of a long article by Castelnuovo and Enriques in the *Encyklopädie der Mathematischen Wissenschaften* of 1914 was the crown of all the research undertaken by the school up to that moment, an official recognition of its significance on the part of the international mathematical community.

The Italian school deserves credit for the definitive re-elaboration of that theory of curves that had been developed above all by German mathematicians, the creation of the theory of surfaces and their complete classification, and the beginning of an analogous construction for algebraic manifolds.

In particular, the theory of surfaces can be considered the point of pride of the Italian school. The first results came from Castelnuovo, beginning in 1891, with the example of a non-ruled irregular algebraic surface, the extension to surfaces of the Riemann–Roch theorem relative to curves and the determination of criteria of rationality. This is where the collaboration with Enriques comes in. The two personalities appear complementary: the volcanic Enriques, who proceeds with extraordinary intuitive power, already almost certain of the outcome which will be reached by successive approximations, less interested in proofs and rigor, an impatient and often distracted reader of the articles of his colleagues, is flanked by Castelnuovo, perhaps less brilliant but who takes it upon himself to state precisely the ingenious intuitions of his brother-in-law and to channel them more correctly and productively. Over the course of 20 years, their collaboration would give birth to a new way of framing the theory of algebraic surfaces that would lead to a sufficiently simple classification, eliminating all special cases. The study of an algebraic surface is reduced to that of a family of curves lying on the surface. Among these, particular attention was devoted to linear systems and continuous nonlinear systems (which exist only on irregular surfaces). In 1914, Enriques set out results that were almost definitive as far as classification is concerned. The surfaces are subdivided into birationally equivalent classes as a function of values assumed from "plurigenera" and from "numeric genus"; in fact, the value of P_{12} alone is sufficient to divide all surfaces into four classes: "the chief problem of the theory of algebraic surfaces is their classification, that is, the effective identification of families of surfaces that are distinguished by birational transformations, each family being characterized by a group of invariant internal characteristics and containing within itself a continuous infinity of classes depending on a certain number of parameters (modules)".

After the war, Castelnuovo devoted himself almost exclusively to probability. In the meantime, Enriques's "star" was equalled – and surpassed, at least at the level of politics – by that of Severi, who also recuperates transcendent methods, with a more marked attention for topological and functional aspects. This is how – in 1928, by then a "grand old man" – Castelnuovo reconstructed the procedures of the Italian school (which, not long after, would be accused of a lack of rigor and excessive condescendence to merely intuitive comprehension):

> Perhaps it is worthwhile to mention what method of work we followed at the time to retrace our steps out of the darkness in which we found ourselves. We had constructed, in an abstract sense of course, a great number of models of surfaces of our space or of higher spaces; and we had arranged these models, if you will, into display cases. One contained the regular surfaces for which everything proceeded as in the best of all possible worlds; the analogy made it possible to transfer to them the most salient properties of the plane curves. But when we tried to verify these properties on the surfaces in the other cases, the irregular ones, the troubles began, and we found exceptions of every kind. In the end the assiduous study of our models had led us to predict some of the properties that had to exist, with appropriate modifications, for the surfaces in both cases; we then put these properties to the test with the construction of new models. If they passed the test, then as the last phase we looked for the logical justification. With this procedure, which is similar to that followed in

the experimental sciences, we were able to establish some traits that distinguished the two families of surfaces.

During the same period, the Italian school of analysis is less monolithic. Here were Dini, Ascoli, Arzelà, Peano and – with the beginning of the new century – Tonelli, Fubini, Vitali, E. E. Levi ... all "great" mathematicians, great names still today known to researchers the world over (and to students as well, thanks to some special results).

Vito Volterra (1860–1940) earned his degree in 1882 and immediately published a paper with the famous example, in the context of the so-called "fundamental theorems of calculus", of a function derivable in an interval whose derivative is limited but not integrable (according to Riemann). He can thus be considered one of the "founding fathers" of functional analysis, with the concept of line function – to indicate a functional (that is, a real number that depends on all values assigned by a function $y(x)$ defined for a given interval or by the configuration of the curve) – and the institution of the relative calculus, up to the development with a Taylor polynomial. He would remind those – Hadamard, and above all Fréchet – who criticized him for a too specific definition of the derivative of a functional (with respect to the subsequent concept of "differential according to Fréchet") that he was not so much concerned with finding the greatest generalization as he was with finding the generalization most adequate for the problem that he was dealing with, a tenet that Enriques emphatically declared had to be followed at all levels of teaching.

Integral equations are the other significant contribution of the beginning of the century, but Volterra was also a mathematical physicist; indeed he would be elected president of the Italian Society for Physics. Actually it is difficult to define Volterra as appertaining strictly to one discipline or another. As a mathematical physicist his principal research regarded the propagation of light in birefringent media, the movements of the earth's poles (today called "the theory of dislocations", but called *distortions* by Volterra), and systems with memory, those which conserve a memory of their history and of which their future state depends on their present as well as past states.

And how can we forget – we've now passed into the 1920s – the pioneering studies in population dynamics? One of these began when Volterra's son-in-law – zoologist Umberto D'Ancona – asked him for a theoretical explanation of a fact that stood out in the statistics regarding fishing in Italian ports of the North Adriatic Sea in the years 1905–1923: that of the total number of fish caught in the years 1915–1918 and the years immediately after, the percentage of large predatory fish caught increased significantly. The exogenous explanation – mainly based on the fact that there had been less fishing during the war – was not able to explain convincingly the different behaviour of prey and predators. Not knowing the contributions of Alfred Lotka, Volterra began studying the problem posed by his son-in-law at the end of 1925. With these words he opened his *Leçons sur la lutte pour la vie*: "I began my research on this topic at the end of 1925, after having spoken with Mr D'Ancona, who asked me if it was possible to find a mathematical way to study the variations in the composition of biological associations". The model analyzed led him to write the system of first-order ordinary differential equations:

$$\begin{cases} x' = ax - bxy \\ y' = -cy + dxy \end{cases},$$

where $x = x(t)$ and $y = y(t)$ represent respectively the evolution over time of the populations of the prey and the predators, and a, b, c, d are positive constants. The model is constructed on the hypothesis, first of all, that an isolated evolution of two species (in terms of the constant percentage rates of their growth x'/x and y'/y) to which is then added the behavioural hypothesis based on the principle of encounters, according to which the effects of predation depend on possible encounters xy in the unit of time. The solution of the system (with the appropriate initial conditions) is set out in explicit form by means of an ingenious method that uses a system of reference to four axes:

$$y^a \cdot e^{-by} = kx^{-c} \cdot e^{dx}.$$

From this solution Volterra derives his three laws that regulate the biological fluctuation in the model: the *law of the periodic cycle* (which proves the endogenous nature of the fluctuations), the *law of conservation of averages*, and finally, that of their *perturbations*, which answers the initial problem. A perturbation due to external causes – for example, the activity of fishing or a change in its intensity – can lead to new average values, and the comparison with the previous values justifies the experimental observation by which a decrease in the activity of fishing in some way favours the smaller species.

Tullio Levi-Civita (1873–1941) was younger than Volterra, but only by 13 years. Those few years, however, were sufficient to project him "wholly" into the twentieth century. Italian mathematical physics is identified with Levi-Civita, above all during the period between the two World Wars. Levi-Civita would become one of the Italian mathematician's most well-known in international circles because of both the importance of his scientific contributions and his extraordinary human and professional qualities.

The son of a civil lawyer who fought with Garibaldi in his youth and later became a politician and senator, by the age of 23 Levi-Civita was already a professor of rational mechanics, and at 25 was given a chair. He taught first in Padua (where he would marry one of his students, Libera Trevisani) and then, from 1918, in Rome. He would become a member of all the Italian academies as well as of the leading academies abroad. Among his principal honours are the Gold Medal of the Accademia dei XL (today the National Academy of Sciences) and the "Royal Prize" of the Accademia dei Lincei awarded in 1907 (together with Federigo Enriques) for his writings as a whole. He was also awarded the Sylvester Medal in London and the Gold Medal of the University of Lima, as well as honorary doctorate degrees from the universities of Amsterdam, Cambridge (USA), La Plata, Lima, Paris, Toulouse and Aachen. In the 1920s and 1930s he was a true ambassador for Italian science.

The dominant figure – along with Volterra – of an entire epoch, he inspired research in every field of rational mechanics and mathematical physics, and was, in the true sense of the word, the Maestro of a great number of students.

Tullio Levi-Civita

His early works brought him to the attention of the scientific world for the solution to the problem of the transformation of the equations of dynamics, which led to the geodesic representation of Riemannian manifolds. Celestial mechanics owes to him the canonical regularization of the differential equations of the three body problem, in proximity to a binary *shock*. In hydrodynamics – an area of research that would occupy him throughout his scientific life – it is sufficient to mention two groups of work that are particularly important. The first regards the theory of drag, begun by Helmholtz to explain the serious contradictions raised by the classic theory of motion of a perfect liquid in which a solid is emerged. The second group is relative to the research of irrotational periodic waves of finite amplitude which propagate without changing form. Stokes and Lord Rayleigh had already grappled with the problem without success (Lord Rayleigh had even come to doubt the existence of this kind of wave); Levi-Civita was instead able to resolve it completely and once again his works on the subject gave rise to numerous other studies (by Struik, Jacotin-Dubreil, Weinstein and others). In mathematical physics we must content ourselves with citing a paper of 1897, in which Levi-Civita deduces Maxwell's equations of the laws of Coulomb, Biot-Savart and F. Neumann, simply substituting retarded potentials in place of ordinary potentials.

These works would already have been sufficient to assure Levi-Civita an important place among Italian mathematicians. But what allowed him to go beyond the restricted circle of specialists was the role he played in the development of absolute differential calculus, which provided numerous applications of various kinds. Set out by Gregorio Ricci-Curbastro starting in 1884, Levi-Civita's professor in Padua, tensor calculus is essentially an analytic theory of differential quadratic forms and their invariants. Levi-Civita was one of the first to understand and show that the new calculus was a powerful instrument for discovery. At the request of Felix Klein, he drafted the celebrated paper published in *Mathematische Annalen*, entitled "Méthodes de calcul différentiel absolu et leurs applications", which made its importance evident. A few years later, Albert Einstein would adopt tensor calculus as the most appropriate instrument for solving the mathematical problems posed by the theory of general relativity.

Immediately after, it was again Levi-Civita who made an ulterior refinement to tensor calculus, with the discovery (in 1917) of the notion of parallel transport:

> ...making the fundamental notions of absolute differential calculus has permitted a theory – up until that time purely analytical – entrance into the realm of geometry. This had profound repercussions in the development of geometry itself, to which Levi-Civita's discovery gave new impetus (comparable to that of 50 years ago with Klein's "Erlangen Programme").

Levi-Civita's paper marks the beginning of a new general theory of connections (elaborated by H. Weyl, J. A. Schouten, O. Veblen, L. Eisenhart and É. Cartan), capable of providing physics with new geometric schemes.

In short, another great mathematician – internationally known – like Castelnuovo, Enriques, Severi and Volterra. But the springtime of Italian mathematics is something more. Something that should be contained – by definition – in the word *mathematics* but which in fact identifies a minor trend, at least looking at the cultural tradition (recent and less recent) of Italy.

Let's go back to Enriques. The decisive turn about in his intellectual and human life occurred in 1906 when he published *Problemi della Scienza*, a fundamental work still quite readable today. A book perhaps a bit "naïve" from the philosophical point of view, it was nevertheless vibrant, and was immediately translated into English, French and German. Enriques (who was 35-years old at the time) had an extremely ambitious project: he tried to construct a general theory of knowledge. Knowledge of the truth, of objective reality, which could be reached above all thanks to scientific research. Essentially, this is reasoning about scientific knowledge. The book immediately became a topic of debate, and enjoyed a wide distribution thanks to the rigor of the arguments, the clarity of the writing, and the variety and vastness of the themes dealt with. During the same years, Benedetto Croce wrote his *Logica*, a most rigorous work at the level of theory, which explicitly negated the value of science on the plane of truth: the value of science is strictly instrumental. Enriques and Severi cannot agree with that; they wrote some political articles about Croce's book and so were castigated in a sarcastic paper entitled "Se parlassero di matematica..." ("If they want to talk about mathematics...").

Aside from the polemics, one fact is evident: Enriques's scientific and cultural project goes beyond mathematics. In the effort to realize it, he moved on all fronts: he participated in the association "Mathesis", he became president of the Italian Philosophical Society, he founded the journal *Scientia* (which would enjoy the collaboration of figures such as Ostwald, Mach, Bergson, Poincaré, Tannery and Pareto), he concerned himself in the National Italian Federation of Teachers, he intervened in the debate concerning the reform of middle schools, he published textbooks for schools in the guise of instruments for a new form of education, he spoke on "the philosophical renaissance of contemporary science" and on "Hegel's metaphysics" considered from a scientific point of view, etc. Even the organization in Bologna of the "International Congress of Philosophy" of 1911 was a very natural fact for Enriques: philosophy is a summit which can be reached (and must be reached) by many routes and *in primis* via scientific knowledge. Philosophy without science is empty, as Einstein more or less said some years later. It is possible then to imagine a faculty of philosophy as the crown of all university studies, with the exception of those that are decidedly professional, such as medical or engineering schools.

The Italian neo-idealists also considered philosophy as the apex of human activity, but it is a discipline for professionals, not for dilettantes such as scientists. The debate between Enriques and Severi on the one hand and Croce and Gentile on the other is not therefore merely academic. It is a struggle between different cultural projects that refer to decidedly different civil projects (which school, which university, which education for young people). Enriques occupied spaces that were academic, cultural and institutional. He looked for channels of communication between the world of scientific research and that of the "other" culture, *civil society*, and his effort inevitably clashed with the project of Italian neo-idealism. The cultural history of Italy and the scholastic experience of all Italians still suffers from his defeat.

Volterra's personality and planning skills are different, but there's no lack of analogies. Volterra never toys with philosophy. Above all he was a man of power, who began his public career with the nomination to senator in 1905. Even his project – that of an outstanding mathematician – goes beyond mathematics. He sees as vital to the mathematical and scientific world the need to project himself beyond his own limits, in order to "export" his own reasoning. The interest is reciprocal: mathematics needs such projections for its development; the other cultures, and even society and the nation's government, need the originality and rigor of the mathematicians. The founding in 1907 of the "Società Italiana per il Progresso delle Scienze" (SIPS, Italian Society for the Progress of the Sciences), of which Volterra is also the first president, likewise pursues a similar dual objective. The internal one regards the scientific community, which must become conscious of the intellectual role proper to it. The specialization of studies is a necessity, but one that should not, however, lead to complete breaks and isolation within closed worlds propelled by a merely technical dimension. This awareness is the necessary premise for putting strong pressure on political powers to overcome their immobility, recognize the social usefulness of science and understand how to find the right collocation for the scientific world. This is the second objective of the SIPS: to participate in the development of the modern nation, one that recognizes the social function of science, as was already the case in more evolved European countries.

The experience of World War I reinforced Volterra's project. From the experience of the war and its outcomes came the idea of a National Centre for Research (CNR) and the Italian Mathematical Union (UMI) to coordinate and direct research and facilitate its use for the nation's progress, including economic progress. Volterra was also the first president of the CNR (in the same years in which he was also president of the "Accademia dei Lincei"). Then Fascism arrived, and Volterra paid for his opposition as an "old" liberal and his support of Croce's 1925 "Manifesto of the Anti-Fascist Intellectuals" by not being reconfirmed as president of the CNR. He was succeeded by Guglielmo Marconi.

Fascism swept away the liberal Italy of Giolitti. It also brought an end to the springtime of Italian mathematics: it is as if the end of the war brought with it the awakening of mathematics from a long, passionate dream and a return to adulthood (?) and sober-minded dedication to its theorems.

Of our protagonists, there are those – like Enriques – who carried on as a mathematician, philosopher, and historian of science, even accepting Gentile's invitation to collaborate on the *Enciclopedia Treccani*, one of the great cultural initiatives of the Fascist regime.

There are those – like Severi – who officially supported Fascism, and emphatically so. At the beginning a Socialist in Padua, and still an anti-Fascist in Rome at the period of Matteotti's assassination and the *Croce manifesto*, Severi made the "great leap" on the occasion of the founding of the "Accademia d'Italia". It was Severi – not Enriques, the expected favourite – who would be the only mathematician to have the honour of bearing the title "Italian academic" (and the accompanying stipend). It was Severi who was one of the great "prompters" of the pledge of allegiance to Fascism in 1931, with a dual aim of ending the disagreements between Fascist and anti-Fascist intellectuals, putting all of them in the same boat, and of identifying the diehard opponents of the regime. Severi was a strong figure. Certainly not a nice one. Think of his responsibility in leading the Italian school of algebraic geometry towards isolation, almost to the point of being completely unaware of what was happening outside; think of his political career: socialist, later Fascist, and later still, after the liberation, no stranger to the "salons" and encounters with exponents of the Communist world. However, we must recognize his attempts – often successful, as for example the founding of the INDAM (the *National Institute for Advanced Mathematics*) – to live mathematics as a protagonist, and to make mathematics itself a protagonist, not reducing it to a purely instrumental role external to every cultural sphere where the nation's future was designed.

There are those instead who opposed Fascism and its politics in the fields of education and culture. One was Castelnuovo, who presided over a commission of the *Accademia dei Lincei* on the educational reform proposed by Gentile and which sounded the alarm as to the dangers this posed to scientific education in Italy.

Castelnuovo sought all possible ways to modify the reform, cultural as well as operational (such as the knotty question of the unification of the chairs of mathematics and physics). During World War II, concerned about the young Jewish men and women who could not attend university because of the racial laws and would be forced to miss years that were critical to their education, by now almost 80-years old, he organized University courses to be held in the Hebrew school (and taught by professors of great prestige, including Enriques), obtaining recognition from the University of Fribourg in Switzerland. The moment Rome was liberated, Castelnuovo immediately concerned himself with obtaining from the Minister, philosopher Guido De Ruggero, Italian recognition for the exams taken by the students at the clandestine university.

We've already spoken of Volterra and his opposition to Fascism. In 1931 he was one of the very few – only about ten! – university professors who refused to swear allegiance to the new regime. The dignity, coherence, and sobriety of his gesture – the last fruit of the springtime of the beginning of the century – are still moving even today, almost 80 years later.

Levi-Civita lived through great difficulties before deciding to swear allegiance. In the end family questions and guardianship of the "school" prevailed, though he would never swallow that bitter pill. Levi-Civita had a different nature. A socialist, a consistent and uncompromising pacifist during the war of 1915–1918, anti-Fascist and, in the eyes and words of the police of the regime, "communist", he never wanted to mix the world of politics with that of scientific research or of his work relationships. For more than 40 years he was one of the most illustrious professors of Italy, attracting students from the world over, helping and encouraging them with inexhaustible patience and generosity. Many received special proof of his kindness. Many enjoyed his hospitality and would never forget his extraordinary personality. A commemoration of Levi-Civita in the 7 March 1942 issue of *Nature* said: "with his death has disappeared a scientist and an Italian who is painful to lose, and not easy to replace".

Guido Castelnuovo

Guido Castelnuovo was born in Venice on 14 August 1865. He studied with Aureliano Faifofer, who directed him towards mathematical studies. He graduated in Padua in 1886 with Veronese and, after having spent a year in Rome with Cremona, he transferred to Turin where – under the influence of Corrado Segre – he published several fundamental works on the theory of algebraic curves. In 1891 he transferred to Rome, where in 1903 he took over the teaching of advanced geometry. After having contributed, along with Enriques (whose sister he married, a fact that led Severi to nickname them "the two brothers-in-law"), to the founding of the theory of algebraic surfaces and the completion of their classification, his scientific interests turned to the calculus of probability, concerning which he published a paper in 1918 that was very important for the development of that field in Italy. Retiring from teaching in 1935, and emarginated by the race laws of 1938, beginning in 1941 he organised the clandestine Jewish University in Rome (where Enriques also taught), thanks to which Jewish students, excluding from Italian universities, were able to earn credits at the University of Fribourg in Switzerland. After the liberation of Italy, Castelnuovo was responsible for the reorganisation of Italian scientific institutions, beginning with the "National Research Centre" (CNR) and the "Accademia dei Lincei", of which he was president until his death, on 27 April 1952. He was nominated Senator for life in 1949.

Guido Castelnuovo

Federigo Enriques

Federigo Enriques was born in Livorno on 5 January 1871. He completed his university studies in Pisa, earning his degree at the "Scuola Normale Superiore" in 1891. In 1892 he came into contact with Guido Castelnuovo, starting to work on algebraic surfaces. After a period spent in Turin with Corrado Segre, he began to teach in Bologna, where he stayed until 1922, when he transferred to Rome. He was president of the "Italian Philosophical Society" and the "Mathesis Association", founder of the journal *Scientia*, and was for a long time director of the *Periodico di Matematica*. He published books for teacher training and scholastic manuals that were used throughout the century (the famous Enriques-Amaldi). He was a section director of the *Enciclopedia Italiana*. Removed from teaching and from all positions by the race laws, he continued to publish abroad, using the pseudonym Adriano Giovannini. He died in Rome on 14 June 1946.

Federigo Enriques

Francesco Severi

Francesco Severi was born in Arezzo on 13 April 1879. Thanks to a scholarship awarded to him by an institution in Arezzo, he was able to study mathematics in Turin with Corrado Segre. He was assistant first to Enriques and then to Bertini in Pisa. In 1922 he transferred to the University of Rome, rising to the position of rector. He wrote more than 400 articles and books, mainly on algebraic geometry, a field in which he introduced new concepts and techniques: among all of those he is credited with, it is sufficient to mention the notion of algebraic equivalence and the theory of series of linear equivalences.

He was considered to be a brilliant speaker and an extraordinary teacher.

After a period in which he first came out against the Fascist regime, he began to grow progressively nearer to Fascism, finally becoming a leading figure. He created the "National Institute for Advanced Mathematics" and was its first president. After the liberation of Italy, he was increasingly excluded by the international mathematics community and some of his results were harshly criticised for lack of rigour. He died in Rome on 8 December 1961.

Verlaine and Poincaré

Something invincible kept me from ever making Verlaine's acquaintance.

I lived very near the Jardin du Luxembourg; it would have taken me only a few steps to reach the marble table where he would sit from eleven to midday, in the back room of a café which took the form, I don't know why, of a rocky grotto.

Verlaine, never alone, could be glimpsed through the windows. The glasses atop the marble table were filled with a green wave that might have been drawn from the cloth of the billiard table, the emerald pool in that nymph's lair.

Neither the allure of fame that was then at its peak, nor my curiosity about a poet whose thousands of musical inventions, delicacies and depths had been so precious to me, not even the appeal of a frightfully uneven career and of a soul both so powerful and mean, ever triumphed over my obscure struggle with myself and a kind of holy terror.

But I watched him go by almost every day, when, upon leaving his grotesque cavern, he would head, gesturing, toward some cheap tavern near the École Polytechnique. That damned man, that blessed man, limping along, would beat the ground with the heavy staff of wanderers and cripples. Pitiable, with flaming eyes under bushy eyebrows, he awed everyone on the street with his brutal majesty and the brilliance of his resonant words. Flanked by his friends, leaning on a woman's arm, he would speak, pounding on the pavement, to his small, devoted retinue. He was given to brusque stops, dedicated to the fury of his raging invective. Then the dispute would move off. Verlaine, along with his followers, would move away, with a painful clattering of his clogs and cudgel, unleashing his magnificent wrath that would sometimes be transformed, as if by a miracle, into a laugh almost as fresh as that of a child.

A few minutes before Verlaine, I rarely missed seeing another passer-by of a completely different nature. He had a curved back, a short beard, clothing that was sensible and serious, a rosette from the Legion of Honor. His gaze, through the trembling crystal of a pince-nez, was empty and fixed. He would walk, vaguely led by his heavy, slanting brow. His uncertain steps seemed to be at the mercy of the most inferior powers of his being. The finger of this illustrious passer-by absent-mindedly drew along the walls that fled away behind him – unconscious arcs that

C. Bartocci et al. (eds.), *Mathematical Lives*,
DOI 10.1007/978-3-642-13606-1_3, © Springer-Verlag Berlin Heidelberg 2011

belied the profound psychological state of a geometer's brain; and the body of his spirit moved as well as it could in our world, which is only one world from among the many possible. The endless internal labor that leads thinkers to enlightenment, fame, and sometimes, with equal indifference, to their death beneath the wheels of a cart, possessed Henri Poincaré.

Regularly moved, like Verlaine, by the law of his table, Poincaré, returning home, would precede Verlaine over the same route. He seemed to me to presage the poet's appearance – by nearly 10 min. I was amused by these meridian transits of stars so dissimilar... I pondered the immensity of the spiritual distance between them. What different images lodged inside those two heads! What incomparably different effects the sight of even the same street could produce in those two systems that followed so quickly one upon the other. In order to conceive of it, I had to choose between two admirable orders of things that were mutually exclusive in appearance, but that resembled one another in the purity and the depth of their purpose. . ..

As for the two passers-by, however, I found they had in common only a similar obedience to the secret summons of midday.

From: P. Valéry, "Passage de Verlaine", from *Études littéraires*.

Bertrand Russell

Paradoxes and Other Enigmas

Gianni Rigamonti

Bertrand Russell lived for almost a hundred years (1872–1970), and was a prolific writer for more than seventy of them, ranging from the foundations of mathematics to logic, from consciousness theory to the history of philosophy, from moral philosophy to political debates. But he is perhaps better known for his pacifism and his militant laicism than for his theoretical work. In 1916, as World War I raged on, his hostility towards the war cost him his position at the university and a period of time in prison; in the last 20 years of his life, he was an active supporter of the anti-nuclear movement.

It was in fact during a protest against nuclear rearmament that I chanced to hear his voice. In 1961 in England, at Easter time (I was 21), there was a wonderful, colourful, cheerful anti-nuclear protest march that wound up at London's Marble Arch, after covering about 70 km, as I recall, in 1 or 2 days. We came from all over Europe. In the final bit we were joined by Russell, then almost 90-years old, who gave one of the speeches at the closing ceremony. To tell the truth, I have to say that I understood all of the speakers except Russell, in spite of the fact that my English was good. I had never before heard, nor have I since, a voice as cavernous as his. Whether it had come to be like that with age, or had always been like that by constitution, I can't say, but I remember that it rained down – along with a copious, warm tepid spring shower – on thousands of wet, reverent and happy young people who didn't understand a word of it (at least, those who were not English).

But here, of course, we are interested in Russell the philosopher of mathematics, not Russell the militant pacifist.

Russell dealt with *Grundlagenforschung* (research on foundations) for a little over 10 years, and after the publication of volume three of *Principia Mathematica* in 1913, he turned his attention to other fields that were less technical and abstract. During that period, brief though it was, he achieved significant results, which I will try to summarise.

In reconstructing the work of a thinker, it is necessary to begin with the state of the art facing the thinker at the moment he began working. In Russell's case, the state of the art was characterised by *Frege's logicism*.

C. Bartocci et al. (eds.), *Mathematical Lives*,
DOI 10.1007/978-3-642-13606-1_4, © Springer-Verlag Berlin Heidelberg 2011

Bertrand Russell

Gottlob Frege (1848–1925), the father of modern logic, was convinced that the whole of Arithmetic[1] could be constructed with purely logical structures, and that it had no *sui generis* principles that were irreducible by pure logic. More precisely, Frege believed that it was possible to define both 0 and, beginning with any natural number n, $n + 1$, and thus all natural numbers, by the sole means of the identity $(=)$,[2] the negation (not), the implication (if), the universal quantifier (in ordinary language, "each" and "all") and the so-called "axiom of comprehension", according to which for every clearly defined predicate P there exists a set of all and only the things that are P; here "clearly defined" means that for every object x, either x is P or x is not P, without any ambiguity. Let's give a couple of examples. If we take human beings as the domain in question, we can see right away that the predicate "only child" is clearly defined (for each of us, having or not having brothers and sisters is absolutely non-ambiguous),[3] while the predicate "nice" is not (when asked if someone "is nice", we can't always give an answer with certainty, and not all of us give the same answer). Consequently, the axiom of comprehension guarantees the existence of the set of only children, but not that of the set of nice people.

This having been established, the natural numbers can be defined (simplifying Frege's meticulous and rigorous procedure for the sake of brevity):

---

[1]But not geometry, and thus, not all of mathematics.

[2]Intended not as a numerical equality but as an ontological coincidence, as *unum et idem esse*.

[3]Knowing whether or not we have brothers or sisters may be more problematic, but here we're not talking about those who know that they don't have brothers or sisters, but only about those who don't have any.

1. 0 is the empty set.[4]
2. Consider the set $0 \cup \{0\}$ obtained by adding to 0 the set $\{0\}$ whose only member is 0. The only member of this set is the empty set 0. We define 1 as the set of the sets that are in one-to-one correspondence to $0 \cup \{0\}$.
3. Let x be a member of 1, and consider the set $x \cup \{x\}$ obtained by adding to x the set $\{x\}$ whose only member is x: each member of $x \cup \{x\}$ is equal to the only member of x, that is, to x itself. We define 2 as the set of the sets that are in one-to-one correspondence to $x \cup \{x\}$.
4. Let's suppose that we have already constructed the natural number n, let x be a member of n, and consider the set $x \cup \{x\}$. Each member of $x \cup \{x\}$ is either equal to a member of x or equal to x itself. We define $n + 1$ as the set of sets that are in one-to-one correspondence to $x \cup \{x\}$.

The intuitive idea behind this construction is that a natural number is a set of equipotent sets. More precisely, n is the set of sets of n members: 2 is the set of doubles, 3 is the set of triples, etc. Except that, set out in this way, the idea is circular. If instead we move by means of the definitions 1–4 stated above, the circularity disappears.[5]

What does all of this have to do with the axiom of comprehension? Each natural number n is the set of all (and only) the equipotent sets of a given set constructed according the procedure described above. Thus, in order for there to be a given n, the totality of the sets must be given, and in order for the totality of the sets to be given, it is indispensable that there be a principle that establishes the sufficient and necessary conditions for the set's existence. This principle is precisely the axiom of comprehension, and thus it is an essential part of Frege's logicism.

But the axiom of comprehension is untenable, and Russell became aware of this in 1902. This is his best known, and perhaps most important, discovery.

The discovery lay in the fact that a contradiction arises from the axiom. In order to see how, let's first define the notions of "abnormal set" and "normal set":

x is an *abnormal set* if and only if it is a set and is a member of itself.

x is an *normal set* if and only if it is a set and is not a member of itself.

For example, the set of poetry is not a poem, and so it is normal; the set of natural numbers is not a natural number, and so it is normal; the set of men is not a man, and so it too is normal. However, the set of infinite sets is infinite, and thus it is abnormal.

The notions of abnormal and normal sets are well defined. In fact, given a set M and an object x, it is self-evident whether or not x is a member of M; thus it is self-evident whether or not M is a member of itself; but this is like saying that it is self-evident whether M is an abnormal or normal set.

---

[4]According to the axiom of comprehension, there will also be the set corresponding to predicates that do not appertain to any objects; for example, the set "grandfathers who never had children" will naturally be empty.

[5]In definitions 1–4 we have also used the set operation of union and the notion of one-to-one correspondence. But it can easily be proven that both are reducible to the basic logical operators mentioned at the beginning of this section.

Thus, being an abnormal (or normal) set is not like being a nice person, but more like being an only child, and the two sets – of abnormal sets and normal sets – exist. Call y the set of normal sets. Now, is y normal or no? For starters, let's try to reason out the hypothesis:

A. y is abnormal.

But "y is abnormal" is equivalent to "is a member of itself": that is, it is a member of y (since it's y), that is, it is a member of the set of sets that are not members of themselves. It follows that it does not belong to itself, or in other words, that it is normal. Hypothesis A is self-destructing. Let's try then with another hypothesis:

B. y is normal.

Now, if y is normal then it is not a member of itself, that is, it is not a member of y, the set of sets that are not members of themselves. Thus it is a set, but it is a member of itself – that is, it is abnormal. Thus, hypothesis B is self-destructing as well. So we find ourselves faced with a contradiction that we can't escape from if we admit that y exists – but it has to exist, because of the axiom of comprehension. In other words, the axiom of comprehension inevitably leads to a contradiction, but that which leads to the contradiction is false. It is not true that every well-defined predicate P corresponds to a set of the things that are P. The axiom of comprehension, however, is the axis supporting all of Frege's logicism, and once it is removed, the reduction of arithmetic to logic, at least in the way Frege attempted to do it, collapses. The discovery of this paradox made Russell famous, but it was tragic for Frege.

Still, Russell himself was a logician who was in some ways even more extreme than Frege, because he didn't consider geometry as descending from an a priori intuition à la Kant, irreducible to pure logic. He was convinced that the whole of mathematics could be constructed starting from logic alone; that the two disciplines, traditionally separated, in reality formed a single system; that, if it were truly necessary to distinguish between the two, then logic might be called the first chapter and mathematics the chapters which followed, of one unified treatise. For Russell, however, this was just a useful subdivision, without any deeper significance.

This unified system was a program to be carried out, and certainly not an existing doctrine. Nor could it be carried out by proceeding along the simplest route, that of the axiom of comprehension. It was necessary to follow another path, which Russell believed he had found in *type theory*. This made it possible for him to surmount the difficulty – which in his opinion absolutely had to be avoided in laying out possible foundation of mathematics – of impredicative definitions.

The notion of impredicativity is so important that it warrants a digression that goes well beyond Bertrand Russell and his theories.

A definition is said to be *impredicative* when it refers to a multiplicity of which the defined object is part. For example, an impredicative definition is one that defines object y as a set of the sets that are not members of themselves, or to put it another way, defines object y as the set of normal sets; in both variants the definition refers to the totality of sets to identify one particular set.

Not all definitions are impredicative. If, for example, I say,

the electoral body of a democratic nation is the set of its citizens who are adult, without a criminal record, and without mental deficiencies or serious psychological disturbances

then I am defining a certain totality, placing conditions on its members and certainly without making reference to a totality of a superior level of which the *definiendum*, the set being defined, is a member.

In everyday language, we constantly use impredicative definitions, but these don't cause any particular problems, nor in general is there any good reason not to do so. We can see this from an example. Let's take a concept such as:

The oldest Italian

This is certainly impredicative, because the person in question is defined on the basis of a larger totality, in this case, of all the Italians. What comes out of this however, is a notion that we can handle just as well as that (undoubtedly predicative, in the sense that it refers to only certain properties of its members) of the electoral body. Just like we don't have any problems stating:

the electoral body is less numerous than the overall population,

we don't have any problems stating that:

the oldest Italian is more than a 100-years old

or that:

no one stays the oldest Italian for long

or again, that:

the oldest Italian is probably a woman.

But while at the level of everyday language no one ever proposed doing away with impredicative definitions, without which we would have to give up many concepts that are clear and certain, in mathematics several scholars – the most illustrious of whom was perhaps Henri Poincaré – recommended abandoning them, in order to abide by the principle that a set must always and only be defined on the basis of some property of its members. The set of even numbers is admissible, and so is that of prime numbers, and that of rectangular trapezoids, but the set y of normal sets is not, because the property of being a normal set immediately poses the question of whether set y, given that it exists,[6] possesses that property or not, while this problem doesn't even arise for the "normal" predicative sets (such as that of prime numbers, or even numbers). And according to Poincaré – among others – this is inadmissible. It is the illegitimate creation of a new member with a given property starting from a previously given totality of things that are P and, at the same time, extend beyond this totality, which is thus no longer defined unequivocally.[7] In order to avoid this kind of mess – say those who think like Poincaré – there has to be a total asymmetry between *definiendum* and *definiens*. What is defined and what does

---

[6]We have already discovered by other means that this is impossible; but here we have to forget that, because we are discussing the notion of impredicativity independent of Russell's paradox.

[7]To clarify further: from Poincaré's point of view, defining a set means introducing a new object as the collection of objects already existing. If being a member of that collection is equivalent to

the defining have to be radically distinct entities, or else the resulting confusion will lead to paradox.

Without a doubt, at this point the situation appears to us to be highly controversial. On the one hand, we have the continuous use of impredicative concepts not only in everyday language but in many mathematical proofs as well; on the other, there is the fact that the idea of a rigorous separation and asymmetry of what is being defined (*definiendum*) and what we use to define it (*definiens*) seems plausible intuitively.

But if we try to go into the problem more deeply, we see that our attitude towards the impredicative is tied to our way of interpreting existence in mathematics. Why, at the level of everyday language, do we unquestioningly accept statements that are clearly impredicative, such as "the oldest Italian"? The reason is, in all likelihood, that we take such a statement as a simple description of an entity that already exists, mediated by a set whose elements are all already given (including the *definiendum*) and which is thus itself completely given. An expression such as "the oldest Italian" does not constitute the object in question. The object in any case exists, on its own, and we use the expression "the oldest Italian" only as a way of saying *oh look, this thing here is thus and so*, where *this thing* exists independent of the fact of its being thus and so, and so any trace of circularity is removed.

But while everyone, apart from a few fringe radical philosophers, takes it for granted that things are this way in the material world, this is not at all the case in mathematics where there is a division which has existed for less than 500 years but whose roots go back more than two thousand, between those who think that mathematical objects exist independent of our knowledge of them – so that we are limited to discovering them only – and those who instead believe that in defining them, we constitute them, and that their existence is reduced fundamentally to being defined. For the former group – that is (to introduce a term that is perhaps illicit but convenient), for the realists – a mathematical object however defined, even impredicatively, can in any case exist as long as its attributes are not contradictory and the impredicativity of the definition does not destroy its existence, in the way, for instance, saying that a certain Mr X is the oldest Italian does not destroy Mr X. But if defining means constituting, then it's another story: for example, *the set of normal sets* is just what this expression says; however, it refers to another totality, the set of all sets, and on the other hand, the totality of the sets refers, as all sets do, to its members, including the set of normal sets. And we can't even say, "okay, the set refers to the member and the member to the set but this only regards the definition, not the objects themselves", because objects exist apart from the definitions only for realists, but not (again, a term that is perhaps illicit but convenient) for the idealists.

---

possessing a certain property, which regards, as we said, objects that already exist, then it is not a new object, and defining impredicatively means violating this distinction.

And now something singular happens. Russell is not at all an idealist in mathematics; he doesn't reduce mathematical objects to the knowledge we have of them, indeed, he's convinced that we don't create them, but discover them. Nevertheless, he rejects impredicativity.

There are two steps in his thinking. The first, more general, is to abandon the axiom of comprehension, indispensable for avoiding paradoxes like the one discussed above and others which were coming to light in those years (around 1895–1905). The second, more specific, was the elimination of impredicative definitions.

That elimination took place by means of so-called *type theory*, which consists in – reducing it to bare bones – a stratification of objects. At level 0 are those objects (we can call them *individuals*[8]) that are not sets. At level 1 are the sets of individuals; at level 2 the sets of objects of level 1; at level 3 the sets of objects at level 2 and so forth. (In truth, Russell not only takes into consideration the level of ontological complexity of his objects, but also the level of complexity that defines them. Let's say we take any two objects from level 1, or the sets of individuals: the definitions of both objects express the necessary and sufficient conditions for each individual to be considered a member, but they can be quite different in terms of complexity, one being extremely long and the other extremely short, for instance.)

Language as well can be stratified correlatively, introducing infinite kinds of variables: for individuals, which we can write using the exponent 0 (x^0, y^0, etc.), for sets of individuals, which we can write using the exponent 1 (x^1, y^1, etc.), and so forth. An atomic formula is well-formed if and only if it is of the form $x^i \in x^{i+1}$ (x^i is a member of x^{i+1})[9] since each set contains only objects of the ontological level immediately below it. Formulae such as $x^i \in x^i$, or such as $x^i \in x^j$ with $j < i$ or $j > i + 1$, are meaningless.

Once this is established, the paradox of the set of normal sets immediately vanishes. It hinges on the expression y as a member of y, or $y \in y$, but in type theory we have to attribute to y an ontological level, and regardless of which level this is, we in any case come up with an expression of the form $y^i \in y^i$, which, as we have just seen, is meaningless. Russell's paradox no longer exists, because the formula that leads to it can no longer be expressed. The same is true, *mutatis mutandis*, for the other paradoxes that came to be discovered in those years.

But the central idea of type theory necessitated an enormous amount of work before it could transformed into a detailed foundation of mathematics, still retaining

---

[8]In *Principia* at least, Russell doesn't worry about establishing which objects are individuals. It suffices him – for reasons that we will see right away – to introduce a stratification.

[9]It is still true that each set is an extension of a property and each property generates a set, even if now the only sets allowed are sets of objects that are homogenous from the point of view of the stratification in types (and this is sufficient, as we shall see, to avoid the paradox). Thus there is no loss of generality in the statement that each atomic formula is of this form: in fact, saying that "such-and-such object has such-and-such property" is, by dint of one-to-one correspondence of property to set, like saying "such-and-such object is a member of such-and-such set".

its logical structure but immune to the weakness that proved to be fatal to Frege. Russell, with the help of Alfred North Whitehead, developed this foundation in the three volumes of the *Principia Mathematica* published between 1910 and 1913.[10] The *Principia* exerted an enormous influence for more than 30 years; then, they were gradually forgotten and today are read by only a small number of scholars. However, the ideas contained in them continue to circulate.

Intuitively, the ontological stratification of type theory appears plausible, but while we have seen that it resolves the paradoxes, is also creates other difficulties.

The first difficulty is a proliferation of objects that are unusable and embarrassing. In fact, Frege's definition of the "natural number" $n + 1$ ($n + 1$ is the set of sets that are equipotent to $x \cup \{x\}$, where x is any member of n) breaks down into an infinite number of definitions of "numbers" of different ontological levels: there will be the (level 2) set of the set of level 1 of $n + 1$ members of level 0, the (level 3) set of the sets of level 2 of $n + 1$ members of level 1, and so forth; nor can this be otherwise, because in type theory there are no sets that are not homogenous from the point of view of the ontological level of their members. Thus, for example, we will never have the 4 *sic et simpliciter* but the 4 of level 2 (with members of level 1), the 4 of level 3 (with members of level 2), and so forth. And we have no idea what to do with this proliferation.

A second difficulty is that once the impredicative definitions have been done away with, many theorems of fundamental importance can no longer be proved, and a foundation of mathematics that eliminates a large part of what it wishes to establish is not acceptable. Russell tried to remedy this situation by introducing three new axioms. The first two are plausible, but the third is much less so. The first of the three, the *axiom of infinity*, says in essence that infinite sets exist; the second, the *axiom of choice*, says that for every family F of non-empty sets F there exists a choice-set S whose members are in one-to-one correspondence to those of F (and further, that every member of S belongs to that of F to which it corresponds).[11] The third, the *axiom of reducibility*, states that given any theorem in type theory regarding objects of any ontological level, there exists an equivalent theorem formulated exclusively in terms of individuals (objects of level 0) and properties of individuals (objects of level 1).

Of the three axioms, the least tenable and most debated is that of reducibility, which is not only without any intuitive evidence to support it, but is tied to an assumption that is openly empiricist-nominalist: that everything that we say, even the most extreme abstractions, can be transformed into a statement about certain basic objects that we consider concrete or simple or immediate; objects, in short,

---

[10] After the publication of *Principia* Russell would no longer deal with logic and the foundations of mathematics. The reason he gave was that the effort had exhausted him, so much so, he claimed, that his capacity to address complicated extractions had been clearly diminished, but it is natural to suspect that by now he was more interested in other questions that were remote from mathematics.

[11] In simple terms, and putting aside the difficulties that arise when we deal with the infinite: we can construct S by taking exactly one element from each member of F, and different elements from different members of F.

that are respectable from the empiricist-nominalist point of view. But this option is philosophically biased, and its insertion into a theory about the foundations of mathematics must inevitably lead to controversy.

In any case, the *Principia* had the immense merit of presenting for the first time a systematic logical foundation of mathematics, detailed and immune to paradox, offering factual proof that this kind of foundation was possible. They diffused optimism among those who were doing research in the foundations. They stimulated many to broaden – possibly even correcting Russell himself – this field of research. They marked – like many other great syntheses, starting with Euclid's *Elements* – the end of an epoch (that of pioneers such as Cantor, Frege, Dedekind or Peano, but also the "crisis of paradoxes") and at the same time, the beginning of another, already more shrewd in terms of methodology than the brilliant but "naïve" research of 20 or 30 years before.

The synthesis was superseded in the space of a generation, partly for the reasons I mentioned above, and partly for two other reasons: the great results of the 1930s regarding undecidability, which opened completely new horizons; and the discovery, again in the 1930s,[12] of formal systems of logic that were much simpler and more intuitive than those of Russell and Whitehead, which were extremely dry and ponderous. In the final analysis, however, the *Principia* provided a very strong impetus to studies that would give rise, some 20 years later, to the "miracle decade" of modern logic.

---

[12]Initially in Germany, the work of Gerhard Gentzen.

Godfrey H. Hardy

A Brilliant Mind

Roberto Lucchetti

Godfrey H. Hardy was born on 7 February 1877 into a family of teachers in Cranleigh, in Surrey. Right away he showed a great intellectual capacity, particularly regarding mathematics: as a small boy he was already "playing" with numbers. He studied in local schools, where he distinguished himself in all subjects, winning numerous prizes, much to his embarrassment, so much so in fact that he said he sometimes gave the wrong answers in order not to have to undergo the torture of the award ceremonies. However, as he says in his book *A Mathematician's Apology*, written in 1940, he also had a strong spirit of competition, and mathematics became his most effective way of outdoing his schoolmates. At the age of 12 he won a scholarship to Winchester College, the best institution in England, at least as far as mathematics is concerned. If there is such a thing as a stereotypical English college of the beginning of the twentieth century, Winchester is the perfect example. A first class education, but probably full of many rigours that must have been difficult to take for a nature as sensitive as Hardy's. One of the things that made him the angriest about Winchester, he later recalled, was that it was forbidden to dedicate any time to practicing the sports he loved, and for which he had a notable talent, especially tennis and cricket.[1] He left Winchester for Trinity College in Cambridge. He was ranked fourth Wrangler in the first part of the famous and infamous Tipos, final examinations so difficult that they were almost cruel. Hardy detested the very idea of the mathematics on which this exam was based: a rigid, extenuating training to solve problems one of whose main consequences was that of stifling any form of imagination that might be used to grapple with problems. But he didn't refuse to take the exams; not only that, he was annoyed at not having been ranked first. In fact, 2 years later, during the second part, perhaps because he had studied even more aggressively, he achieved first place, a fact that earned him a position at the college.

---

[1] At the age of 50, he could still easily beat the tennis player ranked second at Trinity College.

C. Bartocci et al. (eds.), *Mathematical Lives*,
DOI 10.1007/978-3-642-13606-1_5, © Springer-Verlag Berlin Heidelberg 2011

Godfrey H. Hardy

In 1900 he published the first of his more than 300 articles. From 1906 to 1919 he was a lecturer at Trinity College. In 1912 he began to collaborate with John Edensor Littlewood (1885–1977). Hardy and Littlewood would write almost a hundred papers together. It just might be that their collaboration is the most famous and most fruitful in all of mathematics. In spite of this, there appears to be no testimony to how they worked together: Hardy never spoke of it (in contrast to his descriptions of his relationship to Ramanujan).

One morning, at the beginning of 1913, among the numerous letters that he received each morning, he found an envelope covered with Indian postage stamps, hardly the kind he was used to. Opening it, he found a cover letter, which was written in a broken English, and a mixed up mess of theorems: it was the first letter that he received from Srinivasa Ramanujan (1887–1920). Hardy's first reaction was above all annoyance: a list of the stated theorems, none of which came with a proof, a couple of which were well known to specialists but "passed off" as original. Little by little, he must have begun to harbour some doubts, because, after a day spent following his usual unchanging routine, he went to see Littlewood to show him what he had received. Thus began the collaboration between Hardy and a man considered to be one of the greatest geniuses ever to have appeared on the stage of mathematics, limited only by a total lack of education in the field, the fault of his humble origins.

Ramanujan was a simple clerk who lived with his wife on his meagre salary in the city of Madras. He was also a Brahmin, who followed equally closely the rigid religious precepts of his caste, and the advice of his mother. It would then seem impossible that he would accept an invitation to England, but after having discussed the manuscripts he had received with Littlewood, Hardy immediately decided to invite him. But it was Ramanujan's mother who made it possible for him to accept the invitation, saying that in a dream she had seen her son surrounded by Europeans while a goddess warned her not to interfere with his intentions.

Ramanujan arrived in England in 1914, and he and Hardy immediately began a collaboration that was as rich in results as it was anomalous: it should not be forgotten that not only did Ramanujan know little or nothing of modern mathematics, but he had very little use for the kind of rigour that modern mathematicians believe to be indispensable. Although Hardy was aware that he was dealing with genius, he nevertheless was often forced to teach him elementary mathematical rules.

Their collaboration resulted in five works of the highest calibre, but was brought to halt because Ramanujan became seriously ill after about 4 years in England. After having spent some time in hospital, he returned to Madras, where he died of tuberculosis in 1920.

But let's go back to Hardy. In 1914 he publicly sided with those, including Bertrand Russell, who were against the war. However, his was not a position of ideological pacifism; on the contrary, he was convinced that it was wrong to fight Germany because it was a nation whose culture and organisation he admired. Among other things, this made his relations with other colleagues difficult, a fact which drove him to leave Cambridge for Oxford in 1919. In 1926–1928 he was president of the London Mathematical Society. He spent 1928 and 1929 in the United States, in particular in Princeton and at the California Institute of Technology. In 1940 he published *A Mathematician's Apology*; in 1942 he left teaching, which he in any case claimed to hate. He died in 1947.

His obituary in the *Times*, among other things, noted that,

> He personified the idea of the absent-minded professor. But those who formed the idea that he was merely an absent-minded professor would receive a shock in conversation, where he displayed amazing vitality on almost every subject under the sun. ... Outside the schools Hardy was an expert tennis player. He also had a passionate devotion to cricket. Every year he had all the averages at his finger tips. He was interested in the game of chess but was frankly puzzled by something in its nature which seemed to come into contact with his mathematical principles.

Hardy was first of all a brilliant mind, and then a mathematician of notable fame. He had a life that was, all told, a rather fortunate and happy one: he enjoyed a certain economic freedom, due to an adequate stipend, at least for meeting the needs of a single person; he taught only few classes, and had a lot of time free to do with as he pleased; he received immediate recognition for his skills; he had a rich and stimulating circle of friends. Only towards the end of his life, as his strength began to fail him, did he begin to feel a strong sense of sadness.

Hardy also had his quirks. For example, although he was active and fit and was considered to be quite handsome, he allowed practically no one to photograph him, and often covered the mirrors in hotel rooms. But everyone, or almost, is prey to some such idiosyncrasy. The most obvious characteristics that one finds mentioned when reading the memoirs of those who knew him, or even the *Apology*, is perhaps his extreme competitiveness, his extreme need to judge his work and the work of others, who was competent and who was less, which theorems were beautiful, and which were not.

Here, for example, is an oft-quoted phrase that typifies his thinking:

> I still say to myself when I am depressed and find myself forced to listen to pompous and tiresome people, "Well, I have done thing you could never have done, and that is to have collaborated with Littlewood and Ramanujan on something like equal terms".

There is nothing so very strange in this: people are competitive; they love to judge, and even more, they love to argue. Mathematicians are no different in this regard. None of this is very rational, but it is probably inevitable. Charles P. Snow,[2] who wrote a long introduction to the *Apology*, said that "... his precise ranking must be left to the historians of mathematics (though it will be an impossible job since so much of his best work was done in collaboration)...", as though establishing who came up with a fine theorem was more important than the theorem itself. For that matter, I think a much more interesting problem is to try to establish who is the best soccer player in the world[3]: it has the same degree of uselessness as the other one, but at least it's more fun and more popular.

The fact is that exasperated competitiveness leads ultimately to defeat in any case, because sooner or later it leads to unhappiness. Obviously, Hardy was no exception, and this perfectly explains Snow's fine characterisation of the *Apology*:

> That is why *A Mathematician's Apology* is, if read with the textual attention it deserves, a book of haunting sadness. Yes, it is witty and sharp with intellectual high spirits: yes, the crystalline clarity and candour are still there: yes, it is the testament of a creative artist. But it is also, in an understated stoical fashion, a passionate lament for creative powers that used to be and that will never come again.

This brings to mind Graham Greene's review of the *Apology*, in which he says that, alongside Henry James's notebooks, it was "the best account of what it was like to be a creative artist".

One thing in any case I think is beyond doubt: Hardy was certainly one of the mathematicians who *best wrote* mathematics.

---

[2]Physicist, author, politician.

[3]Or perhaps cricket, as Hardy did.

Hardy: From *A Mathematician's Apology*

It will probably be plain by now to what conclusions I am coming; so I will state them at once dogmatically and then elaborate them a little. It is undeniable that a good deal of elementary mathematics – and I use the word "elementary" in the sense in which professional mathematicians use it, in which it includes, for example, a fair working knowledge of the differential and integral calculus – has considerable practical utility. These parts of mathematics are, on the whole, rather dull; they are just the parts which have least aesthetic value. The "real" mathematics of the "real" mathematicians, the mathematics of Fermat and Euler and Gauss and Abel and Riemann, is almost wholly "useless" (and this is as true of "applied" as of "pure" mathematics). It is not possible to justify the life of any genuine professional mathematician on the ground of the "utility" of his work.

But here I must deal with a misconception. It is sometimes suggested that pure mathematicians glory in the uselessness of their work, and make it a boast that it has no practical applications. The imputation is usually based on an incautious saying attributed to Gauss, to the effect that, if mathematics is the queen of the sciences, then the theory of numbers is, because of its supreme uselessness, the queen of mathematics – I have never been able to find an exact quotation. I am sure that Gauss's saying (if indeed it be his) has been rather crudely misinterpreted. If the theory of numbers could be employed for any practical and obviously honourable purpose, if it could be turned directly to the furtherance of human happiness of the relief of human suffering, as physiology and even chemistry can, then surely neither Gauss nor any other mathematician would have been so foolish as to decry or regret such applications. But science works for evil as well as for good (and particularly, of course, in time of war); and both Gauss and lesser mathematicians may be justified in rejoicing that there is one science at any rate, and that their own, whose very remoteness from ordinary human activities should keep it gentle and clean.

Emmy Noether

The Mother of Algebra

Aldo Brigaglia

On 9 September 1932, just a few months after her 50th birthday (on 23 March of that year), in the course of the ninth International Congress of Mathematicians (ICM IX), Emmy Noether danced with Francesco Severi, a Tuscan colleague who was only 3 years her senior.

Although she certainly didn't arouse any particular physical attraction on the part of her dancing partner, who judged her to be "scarcely possessing any feminine attractions; a small and squat figure", Emmy had ample reason to be happy with herself.

If the 26-year-old André Weil would always consider the Zurich congress the most beautiful he had ever attended, highlighted by the magnificent weather at the time and, in addition to evenings of dancing like the one on 9 September, fabulous trips on the lake, this must have been particularly true for Emmy as well, who in that fantastic 1932 seemed finally to have received the recognition she deserved as a great mathematician, the leader of a new school of algebra. Only 2 days before, on 7 September, she had delivered a plenary lecture to a general assembly ("Hyper-complex Systems in Their Relationship to Commutative Algebra and to Number Theory"[1]), perhaps the highest official recognition that she had ever received, a genuine consecration.

A fervid atmosphere of international collaboration reigned over the congress, that is, a strange international atmosphere. The mathematicians were only vaguely aware of what was already being called at the time a "crisis", and the scientists' world seemed to be hanging on to a raft of old-fashioned values that was destined to disappear in a sea that was increasingly storm-tossed. In just a few months, Emmy, like others of her colleagues, would receive an icy letter from the minister for the *Wissenschaften, Kunst und Volksbindung* of the Prussian government notifying her of her dismissal: *Auf Grunde des § 3 des Berufsbeamtentums vom 7 April 1933 entziehe ich ihnen hiermit di Lehrbefugnis an der Universität Göttingen*. A few

---

[1]"Hyperkomplexe Systeme in ihren Bezichungen zur kommutativen Algebra and zur Zahlenthe-orie", *Verhandl. Intern. Mah. – Kongress Zürich* I, 1932, 189–194.

C. Bartocci et al. (eds.), *Mathematical Lives*,
DOI 10.1007/978-3-642-13606-1_6, © Springer-Verlag Berlin Heidelberg 2011

months after that – exactly a year after the Zurich congress, in October 1933 – Emmy would board the steamship *Bremen*, headed for the United States.

Emmy Noether

But during the Zurich congress all the gathering clouds still seemed very far away. After the painful breach of World War I, international relationships were slowly healing and stabilising once again. After two congresses (those of 1920 in Strasbourg and of 1924 in Toronto) excluding German mathematicians, it had only been 4 years (from the Bologna congress of 1928) since the International Congress of Mathematicians had been truly international once more: there were 247 official delegates in Zurich, and 420 participants, for a total of almost 700 people from all over the world, from all political systems, and of all races. In spite of this, André Weil recalled that "he didn't have that unpleasant sensation of being lost in the crowd that would later ruin many conferences". Who would have thought that all of this, achieved through such painstaking effort, would have been destroyed in just a short time? Who would have imagined that many official representatives of German science – Hermann Weyl, representative of the mythical *Vereinigung,* the German mathematical society, or Landau of the Göttingen Academy, or Courant of the University of Göttingen – would soon lose their teaching positions and their countries, and in the case of Landau, probably a suicide, his very life?

No, as Emmy danced with her fascinating Italian colleague she was certainly not oppressed by dark premonitions of her not so distant future. And if Severi wasn't particularly appreciative of her company, even from a mathematical point of view, both of them knew that at that particular congress – unlike that of Bologna just 4 years before – the centre of attention was neither Severi, nor Italian algebraic geometry, but Emmy Noether, her "new" algebra, and her students (the "Noether boys"). Already a large part of German mathematics could be considered as having been taken over by the "new algebraic Word". Both directly through those who had been Emmy's students themselves, and indirectly, the hegemony of the new mathematics certainly went beyond the strict confines of algebra (Artin, Hasse, Brauer, Deuring, Krull, Witt, van der Waerden, who had just published, in 1931, his *Moderne Algebra,* one of the most influential textbooks of the twentieth century). Its influence extended to number theory, to topology (Hopf), to algebraic geometry (van der Waerden again, and Deuring) and, although in a way that was more complex and less direct, to Hermann Weyl, who recalled his mathematical conversations with Emmy in the cold, dirty and damp streets of Göttingen during the winter of 1927–1928. It is perhaps worthwhile also to note that it was precisely through the theory of group representation that the "new" mathematics of Göttingen profoundly influenced the theoretical physicists such as Born and Heisenberg, who also taught at that university.

Emmy's influence rapidly spread throughout the world. In the Soviet Union, Noether had significant contact with and a profound influence on the great topologist Pavel Alexandrov, who had been in Göttingen in 1923. Emmy had then been in Moscow to teach during the cold winter of 1928–1929 and had forged ties with and influenced the Soviet school of algebra of Pontrjagin, Schmidt and above all Kurosh, who can be considered one of her students. In France, all the attention of the young students was focussed on Emmy's German school. Only 2 years later, in 1934, the Bourbaki group was born, the true apostle of the mathematics of structures, "Noetherian mathematics".

But it was above all in the United States that the spread of Emmy's influence was increasingly evident. In Chicago, Albert followed her and developed his studies on algebra; Mac Lane was a doctoral student in Göttingen; Lefschetz had passed through some years earlier; Zariski was immersed in the study of Emmy's algebra through the books by van der Waerden. By training Zariski was an "Italian" algebraic geometer but once he transferred to the United States he finally grasped the importance and meaning of the new methods: "It was a pity that my Italian teachers never told me there was such a tremendous development of the algebra that is connected with algebraic geometry. I only discovered this much later, when I came to the United States". Shortly afterwards, his book *Algebraic Surfaces* would come out, the first to indicate the need to reformulate algebraic geometry through the systematic use of the new algebraic and topological methods.

In spite of Severi's detached and slightly ironic way of recalling that dance in 1932, the star of the congress was Emmy herself, and the geometer from Arezzo was well aware of it.

Indeed, at that very time Severi was engaged in a serious conflict with the most well known of the Noether boys, Bartel Leendert van der Waerden, who has already been mentioned several times, and who since 1926 had been working on revising the foundations of algebraic geometry according to a new point of view, with a work that he would finish only much later, in 1938, and which would lead to a great number of new papers (14 of which were entitled *Zur algebraischen Geometrie*). The young van der Waerden, only 29, took advantage of the Zurich congress to press the mature Severi for answers to questions and requests for explanations about key points, particularly with regard to the concept of the multiplicity of intersections. Severi, pressured by the young student of his unattractive dancing partner, reacted by producing "an impressive quantity of work" throughout the 1930s.[2]

At that time, during the 1932 congress of mathematicians in Zurich, Emmy Noether was seen by many of those present as representing the future of "Mathematics", while Francesco Severi was seen as representing the past. Emmy had delivered her lecture to almost 800 mathematicians from all over the world, describing her latest research to an audience that was largely unprepared to understand it. Describing the lecture, Fröhlich said, "this outlook puts Noether well ahead of her time". She posed questions that paved the way for the use of the methods of cohomology in algebraic number theory, methods that would be properly developed only in the 1950s and 1960s (by Tate, for example).

On this occasion, Emmy adopted the style most appropriate for the great meetings of the international mathematics community: she outlined the essence of the methods which had made Göttingen the centre of the "new algebra" for more than 20 years and which had just the year before been translated for the first time in a book that was internationally acknowledged as being suitable for teaching (that is, van der Waerden's *Moderne Algebra*). She didn't stop there, however; she also outlined a program for future work which, in later years, would prove how far-sighted she had been.

But let's allow Emmy to speak for herself:

> Today I would like to comment on the meaning of the non-commutative for the commutative: in effect, I want to do this in regard to two classic problems that originated in the work of Gauss ... The statement of these problems has continually changed ... and finally take the form of theorems about homomorphisms and the decomposition of algebras, and at the same time, this last formulation makes it possible to extend the theorems to arbitrary Galois fields. At the same time ... I would like to illustrate the principle of the application of non-commutative to the commutative: by means of algebraic theory simple and invariant formulations are sought for what is known about quadratic forms or on cyclic fields, that is, those formulations that depend solely on the structural properties of the algebras. Once these invariant formulations have been obtained – as in the case of the examples mentioned earlier – these facts are applied automatically to arbitrary Galois fields.

It is certainly not our intention within these few lines to provide an in-depth discussion of these topics but behind these words lies hidden an entire new world, at that time completely unexplored, in mathematics: the world of structures.

---

[2]In particular, I cite the article that appeared in the Seminar of the University of Hamburg in 1933, obviously aimed at clarifying the point of view of the Italian school on critical points brought to light by the new German school. Cf. "Über die Grundlagen der algebraische Geometrie", *Abhand. Aus dem math. Sem. der Hamburgischen Universität*, 9, 1933, 335–364.

The structural method and the reasons for its effectiveness are outlined with great clarity and great efficacy (at least for those of us who have been acquainted with these methods since the first years of university): it begins with the historical roots of the great problems that have characterised the history of classical mathematics; there is no desire for change per se. What is being sought is a statement of the problems that is reduced to the essential (one which depends only on the structural properties of the mathematical object in question). If the essence of the problem has truly been grasped and the correct formulation has been chosen, then automatically the structural theory makes it possible to move from the known facts to generalisations, from the known to the unknown.

After this grand research program, which fascinated not only her "boys" but young mathematicians the world over, there followed several pages outlining the first steps already taken in that direction. It takes us into a forest of "ideals", "cross products", "splitting fields", a whole sea of new and unusual terminology in which probably 90% of her listeners were drowning (Weil himself, some years later, spoke of "constructions full of rings, ideals and valuations, in which some of us feel in constant danger of getting lost").

It comes as no surprise then that Severi considered his dancing partner an advocate of the most rigid formalism, in opposition to the intuition of the algebraic geometers, and in particular to Emmy's father, Max Noether.

I beg to disagree with Severi. All during her scientific life, and in her lecture at the Zurich congress as well, Emmy Noether showed herself to be one of the mathematicians of her times most gifted with intuition. Of course, this was not a visual or geometric intuition; but what if not intuition could have allowed Emmy to perceive the outlines of that vast edifice that is modern algebra whose construction had at that time only just begun? It was evident that she had already clearly identified that edifice in 1921, when she published her first work on "ideal theory".

It was extraordinary mathematical intuition that guided Noether through the labyrinth of structures and their properties, enabling her to identify those that were essential and lent themselves to generalisations, those that were mathematically "significant".

The work of classifying the structures (for example, that of classifying the algebras in which she was working, together with Hasse, Brauer and Albert in 1932) went forward in a way that was not unlike that used by the Italian geometers to orient themselves in the jungle of surfaces and algebraic varieties so that these could be classified according to their birational invariants. In order to overcome this complexity, the mathematicians opened the way with what Weil called *éclair d'intuition*, a flash of intuition, and then – only then – painstakingly reconstructed the details of the course taken, subjecting it to a more accurate logical–rational critique.

Emmy did not disdain publishing works in which her capacity to prove had not kept up with the pace of her intuition. Indeed, in that very year, 1932, she had published an erroneous proof of an exact statement. The correct proof would be published only a few months later by her student Deuring.

On the other hand, it had been thanks to an extraordinary kind of intuition that Dedekind (in a paper that came out in the year that Emmy was born, 1882) was able to see clearly that, given the analogous structures of their respective objects of study, that is, integral and polynomial rings, number theory and algebraic geometry had to have a common foundation.

Emmy, who liked to say *Er steht alles schon bei Dedekind* (it's already all in Dedekind), knew this better than anyone, and was developing it, had already connected it to other underlying algebraic structures (the algebras), had indicated how their structural properties, intimately non commutative, could shed light on problems of commutative algebra as well as on those of number theory.

It was a very substantial message to communicate, but in September 1932 only few were able to grasp its profound significance and incorporate it into a project that aimed at rewriting almost all of mathematics, redefining objectives and methods according to Hilbert's indications and the axiomatic method.

Of course, it would be an exaggeration to characterise the mathematics of the twentieth century exclusively as the creation of the course set by the group in Göttingen, and in particular by Hilbert and Noether, and would lead to the under-rating of other, fundamental areas, often only touched on by the mathematics of structure outlined in the 1932 lecture. But even so, that doesn't diminish the grandiosity and effectiveness of the project that was outlined. The "mother of modern algebra" had undoubtedly formed a family whose impact on the development of mathematics is certainly not one that is going to fade into oblivion.

If Emmy's message wasn't understood, this might in small part be due to her skills as a communicator, which were excellent when it came to a small group of followers, but ineffective on occasions such as this one. Severi, certainly referring to her lecture, found her speaking "messy, awkward, a little lispy". Mac Lane also described Emmy's lectures, saying that they were "excellent, both in themselves and because they bear an entirely different character in their excellence. Prof. Noether thinks fast and talks faster. As one listens, one must also think fast – and that is always excellent training".

In short, she had an expository style that was not very suitable to a learned audience, but which was capable of capturing the attention of those who already had some idea of the importance and meaning of what she was talking about. A style that made it necessary that the evident enthusiasm of the speaker be followed by ongoing discussions later in small groups outside the lecture hall.

André Weil was more drastic, saying that her lectures would have been more useful if they had been less disorganised. Some years later, Zariski, after participating in one of Emmy's seminars at Princeton, said,

> She spoke about ideal theory in algebraic number theory ... and a good deal of it was like Chinese to me ... But she was very enthusiastic and I was trying to learn ideal theory so I went faithfully even if I didn't understand everything. Just watching her was fun, and of course, I felt that here is a person who get enthusiastic about algebra, so there is probably a good deal to get enthusiastic about.

In spite of this lack of understanding, the year that was drawing to a close had been full of achievements for Emmy: a truly magical year. Of course, she was not yet a full professor at Göttingen, but only an associate and not even that officially.

According to Kimberling, Emmy was only an "unofficial associate professor", but since at least 1923 she had been assigned to teach algebra and it was possible for her to direct work on doctoral theses. To be sure, Emmy had not succeeded in becoming a member of the local academy of sciences, but that mattered little (the first scientific society that accepted Emmy Noether as a member was the Circolo Matematico of Palermo in 1908).

In 1932, as had been the case by then for at least a decade, all those who wanted to learn about the latest developments in the axiomatic method and made the pilgrimage to Göttingen, made famous by Hilbert, were above all attracted by Emmy's lectures. And in 1932 Emmy, together with her colleague and in part student Emil Artin, had been awarded the Alfred Ackermann–Teubner Memorial Award for the Promotion of Mathematical Sciences. Moreover, her fiftieth birthday had been the occasion for warm-hearted festivities given by the mathematicians of Göttingen. Hasse had dedicated a paper to her in which he provided proofs of many of her intuitive deductions.

But above all, 1932 was the year in which Noether's methods and teachings received definitive affirmation outside of Germany. In 1932 she had undertaken the task, happy and at the same time sad, of preparing the last, unfinished writings of the 23-year-old French logician and algebraist Jacques Herbrand for publication. Herbrand, closely tied to those who would later form the Bourbaki group, died in a mountaineering accident on 27 July 1931. He had spent his last year in Göttingen with Noether. His death meant the loss of one of the greatest mathematical talents at the very moment when his work was reaching its zenith, when he was full of ideas for future research.

Emmy knew that, although only a few years before, in 1928, her ideas had been almost entirely unknown, they by now were rapidly spreading among the young French mathematicians. In her Zurich lecture she mentioned a still unpublished work by Claude Chevalley on which she had had a strong influence, almost as though she were giving a benediction in advance on the future Bourbaki group.

Perhaps her greatest reason to feel satisfied came from the rapid acceptance of van der Waerden's *Moderne Algebra*, much of which had been the fruit of her lectures, as the author himself generously acknowledged. Van der Waerden can perhaps be considered the most promising of the Noether boys, and had followed her since the winter of 1924. The book produced a ripple effect mainly among the young algebraists. Consider, for instance, its effect on mathematicians as prestigious as Garret Birkhoff,

> ...even in 1929 its concepts and methods [of modern algebra] were still considered to have marginal interest as compared with those of analysis in most universities, including Harvard. By exhibiting their mathematical and philosophical unity, and by showing their power as developed by Emmy Noether and her other students (most notably, E. Artin, R. Bruaer and H. Hasse) van der Waerden made "modern algebra" suddenly seem central in mathematics. It is not too much to say that the freshness and enthusiasm of this exposition electrified the mathematical world – especially mathematicians under 30 like myself.

It is worthwhile to note his use of the phrase "freshness and enthusiasm", because the careless or unprepared reader of this book might get the impression that he is reading a cold, formal textbook. This is not the case: the excitement it created lay in the continual discovery of new worlds, in the apparently natural way in which these new worlds shed fresh light on and give new order to problems that appear to be the most disparate and difficult to approach. This excitement might only have been felt by those who were initiated, but it was real nevertheless.

In 1932 Emmy conquered Japan as well: in contemporary with the publication by van der Waerden, *Abstract Algebra* was published in Japan by another one of her boys, Kenjiro Shoda, who had studied with Emma in Göttingen – not the only Japanese to do so – and who went on to become one of the founders of the Mathematical Society of Japan and rector of the University of Osaka for almost 30 years.

On the train for Zurich, Emmy had met up again with her former student Jacob Levitski, by then 28-years old, who had begun teaching algebra at the Hebrew University in Jerusalem in 1931. His lectures gave rise to the flowering of the Israeli school of algebra, which included Amitsur among others.

But 1932 was above all important for the spread of Emmy's ideas throughout the United States. I have already mentioned the effect of *Moderne Algebra* on Birkhoff, but it didn't stop there. At the beginning of 1932, an article by Hasse had appeared in the *Transactions* of the American Mathematical Society summarising the new ideas. Hasse wrote:

> The theory of linear algebras has been greatly extended through the work of American mathematicians. Of late, German mathematicians have become active in this theory. In particular, they have succeeded in obtaining some apparently remarkable results by using the theory of algebraic numbers, ideals and abstract algebra, highly developed in Germany in recent decades. These results do not seem to be as well known in America as they should be on account of their importance. This fact is due, perhaps, to the language difference or to the unavailability of the widely scattered sources.

A significant effort towards unification of languages (by which, of course, we mean not only German and English, but also the various mathematical languages used by the two different schools) which, a few months later, would turn out to be extremely useful, a kind of groundwork in preparation for Artin's, Brauer's and Noether's arrival in America the following year. This paper would be followed by one co-authored by Hasse and Albert some months later introducing a theorem with an unusually long name – the Albert–Hasse–Brauer–Noether theorem – one central to the theory of algebras.

Thus 1932 marked the arrival not only of "Mother Emmy" but also of her beloved brainchild, abstract algebra.

As we know, the following year would be very bitter indeed for Emmy and for the flowering of German mathematics. The loss of her chair, a summer full of doubts and uncertainties, dotted with unpleasant and humiliating episodes (it is said, for example, that at one informal meeting where there was to be a talk by Hasse, someone (Teichmüller?) came dressed in a SA uniform), then the departure in October for the United States, where she was assigned to a university which was below the level she deserved, the women's college of Bryn Mawr. Even in that situation, in which Emmy had the great satisfaction of instructing one of her best

American students at Göttingen, Olga Taussky (to whose memoirs we owe much), she exerted a new and significant influence, becoming one of the points of reference of the new generation of women mathematicians, who found in the great "mother of algebra" a precise and meaningful model.

In the twentieth century, Emmy Noether epitomised not only algebra but "women's mathematics" as well. We could just easily have spoken about "Emmy's girls" as Emmy's boys. Unfortunately, there was no time to found a genuine American school: on 10 April 1935 Emmy died following an operation to remove a tumour.

We are coming to the end of the story, and yet we have only talked about our subject's final years. So in reverse order I'd like to mention the first 50 years of Emmy's life.

Emmy was the eldest daughter of Max Noether, one of the pre-eminent figures of algebraic geometry in the world, who developed (in keeping with the viewpoint of Rudolf Clebsch) the ideas of Riemann concerning geometry. Max was always considered to be a leader in the field of algebraic geometry by the Italian school. Her brother Fritz was also an excellent mathematician (like his sister, he was forced into exile, but went the opposite direction, to the University of Tomsk in the Soviet Union).

Emmy had studied and earned her degree in Erlangen, where she was born, under the direction of Paul Gordan (the "king of invariants"). Even her enrolment in the university was an exceptional event: it may be that without the influence of her father, Emmy wouldn't even have been allowed to enrol. She was the only woman enrolled in mathematics.

As Emiliana Pasca Noether and her husband Gottfried, Emmy's nephew, said, "Hermann Weyl underlined that Emmy was never in her life a rebel. But who can know what her deepest thoughts were in the early years of the 1900s? We will never know for sure and we can only guess. What matters is that she faced the difficulties and persevered in spite of all the silly ideas about women, to become one of the most important mathematicians of her century".

She had stubbornly pursued her studies in Göttingen with Hilbert, and was finally allowed there to obtain her habilitation there only in 1919, after endless discussions in the department and thanks to the decisive influence of Hilbert himself, who had expressed in terms as colourful as they were effective his opposition to discrimination against women: *Meine Herren, der Senat ist doch keine Badeanstalt* (the faculty is not a pool changing room). The topic of her doctoral thesis was the "theory of invariants" which, at the turn of the century, constituted a significant point of contrast between the old way of doing algebra – essentially algorithmically – and the new way – axiomatically, after Hilbert. The theory of invariants was thus her main area, leading almost naturally to the theory of algebras and its applications in arithmetic and theory of group representation.

Of all of her results, I would like to mention only one, presented in the famous 1921 article entitled "Idealtheorie in Ringhereichen": the structural conditions that make factorisation possible in algebraic number fields, thus making possible the extension to any kind of ring that possesses the condition that each ascending chain of ideals is always finite (called in fact "Noetherian rings"). Her techniques also made it possible to construct, by means of the so-called cross products, a large number of central and simple algebras, and the so-called Brauer group, of fundamental importance for the development of group cohomology.

But while this is the area in which Emmy gave wide range to her methods and became the leader of a school, she also left the mark of her genius in every topic that she dealt with. I will only mention the famous Noether's theorem, born in the context of the calculus of variations, which connects the differentiable symmetry of the action of a physical system to a corresponding conservation law. This theorem is fundamental for analytical mechanics and is widely used in quantum physics.

In the same area, Noether's definition of the theory of invariants led her to deal with relativity as well. Regarding invariants, Einstein's statement of 1918 says it loud and clear: "I'm impressed that such things can be understood in such a general way". Yet again, as in all of her efforts, it was her extraordinary capacity to generalise that was so striking.

I will close by mentioning Solomon Lefschetz's recommendation for Emmy's being hired by an American university:

> As the leader of the modern algebra school, she developed in recent Germany the only school worthy of note in the sense, not only of isolated work, but of very distinguished group scientific work. In fact, it is no exaggeration to say that without exception all the better young German mathematicians are her pupils.

He went on to say that if it hadn't been for her race, gender and leftist political opinions (which were actually moderate), she would have been a high-ranking professor in Germany. But Emmy represented everything that the Nazi regime hated. It is no surprise that she was driven out of Germany, but perhaps this makes her all the more likeable.

Carciopholus Romanus

Of all my childhood companions, one figure still remains shadowy, a figure that I have always tried to grasp among the many recollections that have surrendered themselves so sweetly to being entrapped in my pages.

It is Giuseppe, the little monster, son of Rosa Mangialupini (the lupini bean eater). Who ever would have thought that one day I would find Gauss's dream in the shape of a giant lupini bean? The dream of a non-Euclidean geometry, or as I like to think of it, a baroque geometry, a geometry with a horror of the infinite? But just the other day, during one of my weekly visits to Professor Fantappié, holder of the chair of analysis in the Seminar for Higher Mathematics, I came to know a simulacrum much more complex than the shape of a lupini bean: Steiner's Roman surface. This is a fourth-order closed surface of complex variables. It is as curious a shape as I've ever seen, a tuber the size of a stone, with three navels. The German mathematician Steiner was meditating at the Pincian Hill one morning in 1912 [sic], sitting right on one of the benches where, as a lad, I used to read *Les Chants de Maldoror*. Even the geometers left the name the way it was, with the adjective Roman. T.S. Eliot, in the "Song for Simeon", evoked *Roman hyacinths*, "Lord, thy Roman hyacinths are blooming in bowls". And who knows how these two images came to be married in my mind: the hyacinths and this strange mathematical fruit, a fruit from the gardens of the Mediterranean, a kind of odd tomato, a tomato – let it be understood – with three hooks. Think of the mess the fruit growers make today, when they plant one seed inside another, or three seeds tied in one, or when they marry the lily or the rose; imagine a citron with lemon or orange segments inside, or the bizarre things that Redi wrote about to Prince Leopold. Well then, this shape brings to mind Siamese twins, brothers or sisters with a triple knot, triplets of Siamese tomatoes. Professor Conforto, Professor Severi, Professor Fantappié, three luminaries – Severi tall and curly-haired, Fantappié round and short, Conforto thin and medium height – who were all close to me, looking at that shape, seemed moved, as moved as when Linnaeus found out about the *Lacerta faraglionensis*, the blue lizard that only lives on the Faraglioni of Capri, in the smallest habitat on the face of the earth. "This surface", I said, "is as Roman a fruit as the artichoke". But Severi, Conforto and Fantappié instead enumerated all its marvelous properties: four generating circles, three triple poles, an area that could be calculated using rational integrals, and I don't know what other devilry. It felt like I was hearing Linnaeus talk about artichokes: *carciopholus picassianus, carciopholus guttusii, carciopholus piper-nensis aut romanus.* . . . But Steiner's Roman surface, more than having been raised in the *humus* of the Testaccio and the gardens of the Janiculan, more than having been grown in the fertile ferrous earth of the suburbs, seemed to have been worked from the air and light of Rome, like a fine travertine bowl: it was a limestone sponge with three holes, three bashes, three cavities. A form with three humps, a work of Borromini, that's what it looked like. Imagine an elastic sphere squashed by the points of three cones. It was bound to have special acoustic properties, because it really seemed like it was all ears, it looked like an acoustic hunchback that came down from outer space. Even hunchbacks can have really sensitive hearing. . . . Like my friend Giuseppe Mangialupini. He used to run to tell the priest everything that we said. . . .

From: L. Sinisgalli, *Furor mathematicus*, Ponte alle Grazie, Florence, 1995 (first edition, Mondadori, Milan, 1950).

Paul Adrien Maurice Dirac

The Search for Mathematical Beauty

Francesco La Teana

In a 1963 article in *Scientific American*, Dirac said, "It is more important to have beauty in one's equations than to have them fit experiment".[1] In effect, the search for mathematical beauty is the distinctive mark of his work, and led to results that can be compared to those of Newton and Einstein, even though they have also led him into friendless battles within the scientific community. On the other hand, working alone was another dominant characteristic of his life. In his vast scientific output – comprising more than 190 works between articles and books – his name is associated with that of a collaborator in only four cases.

Paul Adrien Maurice Dirac was born on 8 August 1902 in Bristol. Dirac's childhood, with his two brothers and a sister, was marked by the severity of his father, who isolated his family from all social contact and imposed ironclad rules of behaviour. The children, obliged to speak to him only in French, became increasingly reticent. Speaking of his brother, Paul recalled that "if we passed each other on the street, we didn't exchange a word".[2] The father made the brothers attend the Technical College and the Engineering College, even though Paul's brother wanted to study medicine. His brother committed suicide in 1924, and Paul buried himself in problems of physics and mathematics, becoming so taciturn that he never spoke unless he was asked a question, and even then responding in monosyllables. He distanced himself forever from his father. In 1933 he went to receive the Nobel Prize (won together with Schrödinger) accompanied only by his mother, and when his father died in 1935, he wrote to his future wife, "I feel much freer now".[3]

---

[1] P. A. M. Dirac, "The Evolution of the Physicist's Picture of Nature", *Scientific American*, vol. 208, 1963, no. 5, p. 47.

[2] J. Mehra and H. Rechenberg, *The Historical Development of Quantum Theory*, Springer-Verlag, New York, 1982, vol. 4, p. 11

[3] Margit Dirac, "Thinking of my Darling Paul", in B. N. Kursunoglu and E. P. Wigner, eds., *P. A. M. Dirac. Reminiscences about a Great Physicist*, Cambridge, Cambridge University Press, 1987, p. 5.

C. Bartocci et al. (eds.), *Mathematical Lives*,
DOI 10.1007/978-3-642-13606-1_8, © Springer-Verlag Berlin Heidelberg 2011

Paul Dirac

Dirac enjoyed the kind of study he was forced to undertake, personally going into the theory of relativity, and in 1921 he graduated with excellent grades. He wasn't able to find a job, but he was offered the opportunity to further his studies in "applied mathematics" in Bristol from 1921 to 1923, and later to collaborate, or in any case to interact with, Niels Bohr, Robert Oppenheimer, Max Born and others, before becoming a student researcher at Cambridge in the area of quantum studies.

Dirac adopted a lifestyle that was very reserved, made up of 6 days a week of studying and long solitary walks on Sundays in the countryside nearby. For his whole life he was passionate about travel and more than once went round the world.

The majority of his contributions to physics were made during the 8 years from 1925 to 1933: the formalisation and clarification of quantum mechanics, Fermi–Dirac statistics, the relativistic theory of electrons, the quantisation of the electromagnetic field. In 1932 he was named Lucasian Professor of Mathematics at Cambridge, the chair held by Newton. In 1937 he married Margit Wigner, sister of the physicist Eugene Wigner, with whom he had two children. In 1971 the family moved to Florida, where he died on 20 October 1984.

Dirac was without a doubt the greatest English theoretical physicist of the twentieth century. In 1995, on the occasion of the celebration of his activities in

London, a commemorative plaque was placed in Westminster Abbey (near those of Newton and Maxwell).

Dirac was involved almost exclusively with physics and mathematics, and had no interest in art, music, politics or social activities; he carefully avoided any problems from the outside world that might have led to a change in his lifestyle. When the Manhattan Project for construction of the atomic bomb got underway and a talented group of physicists from all over the world moved to Los Alamos, Dirac refused to take part, even if, in his own way – that is, without changing any of his personal habits – he participated in the effort with some works on a centrifuge capable of separating mixtures of isotopes.

Quantum Mechanics

In 1925 Heisenberg published his famous article on what would become matrix mechanics, noting how the multiplication of two quantum magnitudes $x \cdot y$ was generally different from $y \cdot x$. In a very short time Dirac became convinced that this was the most interesting and important aspect. He translated Heisenberg's theory into a Hamiltonian scheme (which he favoured over all others), reformulating it with Poisson brackets, and thus derived the fundamental equation of motion.

Dirac called his method the "algebra of q-numbers" (where q stands for quantum), that is, the algebra of numbers that don't obey common laws of multiplication, and of c-numbers (where c stands for classic) which make up the q numbers and obey the commutative law. The scheme was analogous to that developed by Heisenberg, Born and Jordan in which the matrices (q-numbers) represented the position and impulse of the electron and were composed of c-numbers (the amplitudes and frequencies in Fourier series). At the time, Dirac was convinced that the algebra of q-numbers would more general and powerful than that of matrices, but with the exception of van Vleck, no other physicist adopted Dirac's scheme, which was judged to be too difficult, preferring instead that of Schrödinger, which appeared in 1926. Dirac himself, on the other hand, carefully studied wave mechanics, applying it (in August 1926) to electrons, proving that these had to obey the so-called Fermi–Dirac statistics, while photons followed Bose–Einstein statistics.

In December 1926, Dirac set out a general quantum description – known as the "theory of transformations" – unifying the three separate formulation of matrices, waves and operators, thanks to the introduction of the famous "Dirac delta function". The delta function "has proved to be of extreme importance in virtually all branches of physics. In the realm of pure mathematics it may be seen as a predecessor of the theory of distributions created in 1945 by the Swiss mathematician Laurent Schwartz".[4]

---

[4]H. Kragh, *Dirac. A Scientific Biography*. Cambridge, Cambridge University Press, 1990, p. 41.

The theory, along with its unified formalism, was published in the volume *The Principles of Quantum Mechanics*, the first edition of which appeared in summer 1930. The work reflected Dirac's taste for the abstract and elegant, as well as his dry, uncluttered style and became universally considered to be the classic text on quantum mechanics, an essential component of the library of all students and researchers. The final touches to the theory's formal elegance were introduced by Dirac with vector notations *bra* and *ket*, which he invented and inserted in the third edition of 1947.

By the end of 1926, Dirac's international reputation was firmly established. His works – although superior in terms of the capacity for generalisation, formal beauty and mathematical creativity – were however stigmatised by their having been obtained in concomitance with or following those of other scientists (Born and Jordan regarding the fundamental equation of motion, Fermi regarding statistics, Jordan regarding the theory of transformations). In particular, Dirac "felt that he lived in the shadow of Heisenberg and the other German theorists",[5] even though those were in constant contact and collaboration with each other, while Dirac worked alone.

The Foundations of Quantum Field Theory

In quantum field theory, Dirac was the founder and font of inspiration for fundamental developments which took place at the end of the 1940s, though, strangely, he never accepted that attribution.

The problem was that of the interaction between radiation and matter. The old theory of Bohr explained the emission and absorption of radiation by means of the image – little beloved – of the electron that jumped from one level of energy to another, but which left the phase of interaction unexplained. The use of perturbation methods in quantum mechanics made it possible to deal with the transitions induced by an external field, but did not provide any description of the disappearance of a photon, nor of spontaneous emission. In order to obtain this, it was necessary to develop a quantum theory of electrodynamics in which the forces were propagated with the finite speed of light rather than instantaneously. Dirac addressed the problem in February 1927, with one of his most brilliant ideas. He wrote the radiation field in Fourier series, dealing with the components of the electric field and the corresponding phases – up to that time considered to be c-numbers – as q-numbers. In this way even the vector potential became a quantised q-number ("second quantisation").

All the same, the significant problem of some divergent integrals soon manifested itself in connection with both the energy and the charge of the electron as well as with its formulation as a point-like object. In 1932, in collaboration with

---

[5]Ibid., p. 48.

Fock and Podolsky, Dirac then formulated an invariant relativistic theory which, however, left the problem of infinites unaltered.

Quantum electrodynamics faced a crisis that would be overcome only at the end of the 1940s, thanks to the so-called "renormalisation techniques". Schwinger, Tomonaga and Feynman openly declared Dirac's importance and influence on their work, but Dirac never accepted the developments of the theory, maintaining that rules of renormalisation "is just a set of working rules, and not a complete dynamical theory at all".[6]

Electron Theory

In January 1928, Dirac made his most important contribution to physics, that of the quantum-relativistic electron theory, aimed at explaining the behaviour of electrons that moved at speeds that were relativistically relevant. Before Dirac, in 1926, Oscar Klein had published the so-called Klein–Gordon equation – which Dirac never accepted because it was second-degree (while quantum mechanics required a linear equation) – which furnished negative results regarding the probability of finding an electron in a given place and did not take into account the existence of the electron's spin.

Dirac thus attempted to find a method for making the equation linear. Everything depended on the linearisation of a quadratic expression placed under the radical symbol. Dirac obtained the solution thanks to the insertion of four 4×4 matrices ("Dirac matrices"), arriving at the famous "Dirac equation". From a methodological point of view, "Dirac reduced a mathematical problem to a physical one, and the mathematics forced him to accept the use of 4×4 matrices as coefficients. This again forced him to accept a four-component wave function $\psi = (\psi_1, \psi_2, \psi_3, \psi_4)$... Though logical enough, this was a bold proposal since there was no physical justification for the two extra components".[7]

Dirac's theory proved Sommerfeld's formula for the fine structure of hydrogen and predicted the existence of spin, which up to that time had always been introduced by means of hypotheses that were more or less ad hoc. The magnitude ψ turned out to be a new mathematical object called a spinor, and was later studied by von Neumann, Weyl and others.

Controversy broke out concerning the physical interpretation of the four components ψ_i. Two of the components had been associated with states of positive energy (related to the two values of "spin up" and "spin down"), while the other two represented states of negative energy. Classically, these would have been discarded since they are physically impossible. In terms of quantum theory, such a solution

---

[6]P.A.M. Dirac, "The inadequacies of Quantum Theory", in B. N. Kursunoglu and E. P. Wigner, op. cit., p. 196.
[7]H. Kragh, Op. Cit., p. 59.

could not be adopted because there was the probability of quantum leaps from states of positive energy to states of negative energy. In 1930 Dirac proposed this interpretation: in normal conditions, all of the states of negative energy are occupied by elections and there is no tangible physical manifestation; when radiation is absorbed, an electron can pass from a state of negative energy to one of positive energy. This leap has the effect of creating a free electron, and a gap in negative energy, which manifests itself as a particle that is positively charged, identified as a proton. Since the theory arrived at a contrast between the mass of the electron and that of the proton, Dirac was soon convinced that the gap was a new kind of particle, one that was unknown in experimental physics, which had the same mass and a charge opposite to that of the electron,[8] which he called an "anti-electron". Two years later American physicist Carl Anderson discovered traces of the anti-electron (the positron) in cosmic radiation.

From the point of view of his research methodology, this was Dirac's greatest success: that he had seen in advance on the basis of theory the existence of an elementary component, one that is now there for all to see in today's techniques of positron emission tomography (PET).

The Method of Mathematical Beauty

In the 1963 *Scientific American* article cited earlier, in which Dirac discussed possible future developments in physics, he proposed improving the capacity for investigation by raising the "method of mathematical beauty" to the status of a guiding principle, as it had always been in his own research:

> A great deal of my work is just playing with equations and seeing what they give. Second quantization I know came out from playing with equations. I don't suppose that applies so much to other physicists; I think it's a peculiarity of myself that I like to play about with equations, just looking for beautiful mathematical relations which maybe don't have any physical meaning at all.[9]

According to Dirac, there are two strategies for studying nature: the *experimental method*, which, beginning with observed facts, looks for the relationships that exist between them; and the *method of mathematical reasoning*, which only involves the search for mathematical beauty, the physical significance of which is investigated only later. The first was the method used by Heisenberg in 1925; the second, that of Schrödinger, who he said had found his equation simply by looking for one that was mathematically beautiful.[10] For Dirac, the second method was the

---

[8]See P. A. M. Dirac, "Quantized Singularities in the Electromagnetic Field", *Proc. of the Royal Soc. of London,* A133, 1931, p. 61.

[9]T. S. Kuhn, "Interview with P. A. M. Dirac, Session III", 7 May 1963, in *Archive for the History of Quantum Physics*, p. 15.

[10]P. A. M. Dirac, "The Evolution of the Physicist's Picture of Nature", *Scientific American,* vol. 208, no. 5, 1963, p. 45.

more profitable because nature manifests itself in terms of beautiful mathematical equations. He said that one of the fundamental characteristics of nature appeared to be the fact that fundamental physical laws are described by mathematical theories of great beauty and power, and that a high level of mathematics is required to understand them.[11]

If we try to understand exactly what Dirac meant by "mathematically beautiful", we find that the concept was never adequately clarified, because mathematical beauty "is a quality which cannot be defined, any more than beauty in art can be defined, but which people who study mathematics usually have no difficulty in appreciating".[12] In addition to that of Schrödinger, examples of beautiful theories were those of Hamiltonian formalism and the theory of relativity.

In any case, for Dirac mathematical beauty was not linked to simplicity, nor to complexity, nor to formal rigour.

According to him, simplicity was subordinate to beauty: it often happens that simplicity and beauty are equally necessary, but when they oppose each other, then beauty must prevail.[13] As far as complexity is concerned, to quote Kragh, "if a physical theory, such as the Heisenberg–Pauli quantum electrodynamics, was only expressible with a very complicated mathematical scheme, this was reason enough to distrust the theory".[14] Finally, with regard to mathematical rigour, Dirac maintained that the correct course for future progress went in the direction of not putting excessive effort into the search for mathematical rigour, but rather to find methods that function in practical examples.[15]

This last expresses his pragmatic outlook. Dirac was one of the founding fathers of the physics of the infinitely small and one of the greatest physicists of the twentieth century. His great skill was that of introducing and creating new mathematical methods capable of solving his problems: from q-numbers to readapted Poisson brackets, to the delta function, to the *bra* and *ket* vectors. He was often criticised by purists for the excessive formal liberty he took, but his aim was that of creating, time after time, the mathematics he needed in order to solve specific problems. He didn't hesitate unduly if he had to sacrifice formal rigour, because his goal was to achieve a result that was mathematically beautiful.

---

[11]Cf. "The Evolution of the Physicist's Picture of Nature", *Scientific American,* vol. 208, no. 5, 1963, p. 45.

[12]P. A. M. Dirac, "The Relation between Mathematics and Physics", *Proc. of the Royal Soc. of Edinburgh,* 59, 1939, p. 123.

[13]Ibid.

[14]H. Kragh, op. cit., p. 279.

[15]P. A. M. Dirac, "The Relation of Classical to Quantum Mechanics", in *Proc. of the Second Canadian Mathematical Congress,* Vancouver 1949, University of Toronto Press, 1951, p. 12.

A Monosyllabic Interview

Here is an interview that Dirac gave to a playful journalist of the newspaper *Wisconsin State Journal*, during his first visit to the United States in 1929.

Then we sat down and the interview began.

"Professor," says I, "I notice you have quite a few letters in front of your last name. Do they stand for anything in particular?"

"No," says he.

"You mean I can write my own ticket?"

"Yes," says he.

"Will it be all right if I say that P.A.M. stands for Poincaré Aloysius Mussolini?"

"Yes," says he.

"Fine," says I, "We are getting along great! Now doctor will you give me in a few words the low-down on all your investigations?"

"No," says he.

"Good," says I. "Will it be all right if I put it this way: 'Professor Dirac solves all the problems of mathematical physics, but is unable to find a better way of figuring out Babe Ruth's batting average'?"

"Yes," says he.

"What do you like best in America?", says I.

"Potatoes," says he.

"Same here," says I. "What is your favorite sport?"

"Chinese chess," says he.

That knocked me cold! It was sure a new one on me! Then I went on: "Do you go to the movies?"

"Yes," says he.

"When?", says I.

"In 1920 – perhaps also in 1930," says he.

"Do you like to read the Sunday comics?"

"Yes," says he, warming up a bit more than usual.

"This is the most important thing yet, doctor," says I. "It shows that me and you are more alike than I thought. And now I want to ask you something more: They tell me that you and Einstein are the only two real sure-enough high-brows and the only ones who can really understand each other. I won't ask you if this is straight stuff for I know you are too modest to admit it. But I want to know this: Do you ever run across a fellow that even you can't understand?"

"Yes," says he.

"This well make a great reading for the boys down at the office," says I. "Do you mind releasing to me who he is?"

"Weyl," says he.

The interview came rapidly to a halt. Dirac had said twenty words in all, eight of them monosyllables.[16]

---

[16]H. Kragh, *Dirac. A Scientific Biography*. Cambridge, Cambridge University Press, 1990, p. 72–73.

The Theoretical Intelligence and the Practical Vision of John von Neumann

Roberto Lucchetti

János Lájos Neumann was born in Budapest on 28 December 1903, firstborn of Miksa and Margit, important members of Budapest's Jewish community. His mother's family was well-to-do; his father, a lawyer, was the director of one of the most important banks of the capital. The family's position was similar to that many other wealthy Jewish families in the city, and promised János a life free from economic problems in a cultural environment rich in stimuli. In 1913 his father was even granted a title of nobility, which though he never used, was constantly vaunted by his son.

From the time he was a small boy János lived in an environment that was very effervescent culturally, and he received a first-class education. At first he was privately tutored, then entered one of the best high schools in the city, but his mathematical education continued at the hands of private tutors. The world of Hungarian mathematics was particularly vivacious, sustained by leading lights such as Fejér, Haar and Riesz. Although he had an evident gift for mathematics, his education was also rich in philosophical and scientific-technological subjects. At home his father often spoke of his professional activities, extending the conversation to theories of economics and finance: without a doubt, the environment and atmosphere of his home profoundly influenced young von Neumann's leanings. Throughout his life he was interested in a wide variety of topics, leading to his being not only a talented mathematician who concentrated on its own discipline, but a man of multifaceted scientific interests.

At the beginning of the 1920s, the destruction of World War I weighed heavily on the world of mathematics, and more in general, on all aspects, cultural or not, of life in Budapest and all of Hungary. This led to the migration of almost all the famous, established mathematicians, as well as the youngest emerging talents. Von Neumann was no exception.

In 1921 he enrolled in the University of Budapest, but at the same time he also attended that of Berlin, where he took courses in chemistry and statistical mechanics. In 1922, at only 18, he published his first article in mathematics in collaboration with Fekete, one of his professors. Then he studied chemical engineering at the Zürich Polytechnic, perhaps more to please his father than out of genuine interest.

C. Bartocci et al. (eds.), *Mathematical Lives*,
DOI 10.1007/978-3-642-13606-1_9, © Springer-Verlag Berlin Heidelberg 2011

In any case, these studies likely influenced his lasting interest in mathematical applications. In 1925 he defended his doctorate thesis at the University of Budapest, on the topic of the axiomatisation of set theory, which he had discussed several times with Fraenkel.

Von Neumann

While still quite young von Neumann took a precise position in the debate that animated the mathematics of these times: on one side the logicians, represented by B. Russell and aiming at perfecting classical logic; on the other side the intuitionists, with Brouwer; on a third front the champions of the axiomatic method, with Hilbert. Von Neumann would side with these last, holding that logicians and intuitionists, in spite of their having obtained significant results, had a vision that was destructive for mathematics. The question of axiomatics would interest him for the rest of his life, and would be brought into all the various branches, including the most applied, worked on by John (as he was called by now).

He began to travel to Göttingen, home to perhaps the most prestigious school of mathematics in the world, founded and animated by F. Klein and D. Hilbert. It was here that von Neumann came to know Hilbert, and immediately became a member of his group. He won a grant from the Rockefeller Foundation and was named *Privatdozent* at the University of Berlin. He collaborated with Hilbert on the question of the axiomatic foundations of quantum mechanics, developing among others a general theory of the inner product (or pre-Hilbert) linear operators. His researches would later be gathered in the volume *Mathematical Foundations of Quantum Mechanics* published in 1932.

The beginning of his career was made possible by his family's economic support, since those holding the title of *Privatdozent* did not receive a salary, although they did receive proceeds from the taxes paid by the students. Perhaps partly for this reason, in 1930 he accepted the position of invited Professor in mathematical physics at Princeton, where he taught quantum statistics, mathematical physics and hydrodynamics. Thus began his experience in the United States, although until 1933 he returned regularly to Germany. In September 1930 he participated at the Congress on the epistemology of the exact sciences held in Königsberg, presenting a paper in favour of the Hilbertian method. But that Congress today is remembered for another lecture of no small significance: K. Gödel announced the first version of his celebrated incompleteness theorem, which he would perfect over the next few years. It appears that the significance of Gödel's result was not immediately grasped by those present, with the exception of von Neumann, who discussed it with Gödel himself, and then shortly later wrote to him that he had proven – as a consequence of the incompleteness theorem – that the coherence of arithmetic was impossible to demonstrate.

Gödel's theorem has a strong emotional impact on von Neumann, who from that moment abandoned his research in mathematical logic. Here is how he described, in an article of 1947, his reaction at the time:

> I have told the story of this controversy [about the foundations of mathematics] in such detail, because I think that it constitutes the best caution against taking the immovable rigor of mathematics too much for granted. This happened in our lifetime, and I know myself how humiliatingly easily my own views regarding the absolute mathematical truth changed during this episode, and how they changed three times in succession!

In any case, although the dream of giving mathematics indisputable and coherent foundations vanished, von Neumann believed that the classic way of doing mathematics should not be abandoned (to follow, for example, logical intuitionism) and that the axiomatic method would be a core instrument: even his most applied work would reveal this thinking.

In 1933 Hitler came to power. The experience of Göttingen came to an end because of the numerous dismissals and resignations, and thus began von Neumann's detachment from Europe.

In 1930 he had married Marietta Kovesi, converting to Catholicism on the occasion. In 1935 their daughter Marina was born, but they divorced in 1936. In the meantime, he was named professor at Princeton's Institute for Advanced Study, where over the years he would teach courses in measure theory, operator theory, and lattice theory. He took American citizenship and began his collaboration with the Ballistic Research Laboratory, a laboratory of the American armed forces.

Von Neumann was, however, interested in other important areas. Intrigued since his youth by questions of economics, he had no qualms about harshly criticizing the Walrasian model of equilibrium and began to think about a new approach to economic theory. More generally, his conviction that social behaviour should be guided by rational analysis and studied with mathematical methods led him to game theory, which will be discussed in a later chapter. Here we will only mention the fact that von Neumann was the first to develop the theory of zero sum games, arriving at the formulation of the minimax theorem, and later published – in collaboration with the Austrian economist Oskar Morgenstern – the book *Theory of Games and Economic Behaviour*, today considered the official birth of game theory.

The relatively quiet period at Princeton came to an end on the eve of World War II; the year 1938 signalled a radical turnabout in von Neumann's scientific activity. Military research began to absorb increasing amounts of his time. He worked on detonation waves, the impact effects of explosives, and the impact of projectiles, and spent a semester in an English laboratory where he worked on gas dynamics. It was during this period that he began to take an active interest in problems of automatic calculation. In September 1943 he began to collaborate with the laboratory in Los Alamos, where the top-secret Manhattan Project was developing, its objective the construction of the atomic bomb.

After the end of the war, scientists were divided about the morality of continuing research on nuclear weapons. There were those, such as Oppenheimer, who actively participated in the formulation of an American nuclear policy; others, such as Einstein, sided with the pacifists. Von Neumann repeatedly claimed that his skills were exclusively technical, but he clearly distanced himself from the pacifist movement. His attitudes left no doubt about the fact that he was in favour of the ulterior development of nuclear technology. Whether that was due to his strong hostility first towards Nazism, and later towards communism, or to a kind of patriotism for his adopted country, is still not clear today. In any case, his responsibility in various political and military projects grew, as did his consulting work with different laboratories and research institutes. He reached the apex of his career when he was named a member of the AEC, the Atomic Energy Commission.

His frequenting the most important military laboratories, his defence of the legitimacy of nuclear testing, and his ideas about preventative war have lent credence to the image of von Neumann as a hawk. Many were inclined to see a portrait of von Neumann himself in the mad scientist of the famous film *Dr. Strangelove*. It's difficult to form a thoughtful judgment about his positions. Certainly, his personal life, the tragedy of Nazism, and his conviction that the weaknesses shown by Western democracies had led to World War II, profoundly influenced his thinking. But then, the image of a man thirsty for power and at the same time submissive to the powers that be, addicted to high-speed driving, and obsessed with sex – sex enters into everything, even in mathematics – reveals a kind of ideological intolerance. Perhaps closer to the truth is the image of a man accustomed to being interested and involved in social, political, and economic questions since his youth, which led to his thinking of the scientist as a man, not closed in an ivory tower, but on the contrary, at the centre of problems – of all problems, political and social. And as a scientist, he carried out his work in a manner coherent with this idea.

But let's go back to his scientific interests in order to mention briefly another field in which his genius expressed itself. It is clear that the complex problems that von Neumann always dealt with, from questions of game theory to those more properly military, had brought to his attention the problem of performing long and difficult calculations, which could not be tackled by the human mind in acceptably short amounts of time. On the other hand, the first machines for fast calculation had just begun to be built, so it is not surprising to discover that von Neumann took particular interest in this field as well. And as always, on the other hand, his interest was not focussed on only one aspect of the question. In fact, he worked as much on technical questions regarding the actual construction of powerful calculating machines, as he did on the theoretical foundations of the structure of those machines. Thus he worked, especially with Goldstine, on developing a theory on the principles of the computer. Above all the machine he had in mind had to be designed for applications in scientific research. Thanks to his influence, the Institute for Advanced Study decided to realise his project, with von Neumann following even in the most practical phases of construction (from the search for funding to the building of the hardware). With Goldstine and Burks, von Neumann worked out the logical design for the computer, producing a report in which they described what we would today call "von Neumann architecture". His contributions to the development of numerical analysis were also fundamental, specifically in connection with the use of the computer to resolve complex problems.

The program for the realisation of the project required much more time than had been foreseen and by the time the computer was donated to the University of Princeton in 1957, it was already technologically obsolete. Be that as it may, today von Neumann's contribution is recognized as fundamental for the beginning of computer studies in the United States, even if the directions taken later by computer science may not be those that he had in mind. One of the first, spectacular, applications that von Neumann expected from such a machine was in the field of meteorological forecasts. Perhaps – in his daughter's opinion – he would have expected greater applications in the area of game theory. His daughter also believes that an application which her father never dreamed of, but which he would have enjoyed immensely, was the videogame!

In parallel to his projects in computer science, von Neumann also dealt specifically with the theory of information and the theory of automata. The basic idea was that in some way the computer had to mimic the characteristics of the human brain. And to be able to design such a machine, it is necessary to understand better how the brain functions and to provide a logical-mathematical basis – not only descriptive – for its functions. Von Neumann therefore entered the world of biomedical engineers and neurophysiologists, participating in numerous conferences and explaining his ideas on various occasions. His most celebrated contribution in this field is the book, published posthumously and unfinished, entitled *The Computer and the Brain*, containing the text of some lectures that he had been invited to give at Yale University.

Although he held an important list of offices in public and private institutes, and consultancy contracts with various companies, more than once von Neumann made clear his intention to return to a more academic life. However, his projects would be brutally brought to a halt by the disease that began to afflict him: he was diagnosed with bone cancer. Although he was ill, he continued to work feverishly. But by the end of 1955 lesions to his spinal cord made it difficult for him to walk. In spite of this, in March 1956 he signed a contract with the University of California as consultant to the various departments. But the cancer was by then out of control, and von Neumann died in Washington on 8 February 1957, at the age of 53.

It is out of the question to try to sum up von Neumann's contributions to science in general and to mathematics in particular. This would take much more space and a different kind of competence; it is not a task for the simply curious. What we can say to conclude this brief account of some of the aspects of his life, is that von Neumann was certainly a true giant of the twentieth century, a figure more unique than rare in his astonishing capacity to join a theoretical intelligence of extraordinary depth to a very concrete view of science, to a concept of life that led him to be an extremely important figure in political and social spheres, perhaps a unique example – at least at this level – among mathematicians.

Kurt Gödel

Completeness and Incompleteness

Piergiorgio Odifreddi

On 22 June 1936, while he was going up the steps of the University of Vienna, Moritz Schlick was accosted by a student who first rebuked him for having written an essay that he disagreed with, and then shot him to death with a pistol. At the trial the assassin was declared insane, but after the Nazis annexed Austria in 1938, he was cleared of the charges because he had made himself useful to the system by eliminating a Jewish professor.

In the eyes of the insane and the Nazis, the real crime committed by Schlick – who, for what it is worth, was not Jewish, but was rather a descendant of Prussian nobility – was that he had founded, in 1924, and been the driving force behind the Vienna Circle, that famous congregation of philosophers and epistemologists who met every Friday evening, were inspired by Wittgenstein's *Tractatus* and worshipped logic as much as they loathed metaphysics.

In particular, the members of the Circle believed not only that metaphysics was false, but that it was literally foolish. This opinion of theirs was expressed in 1931 by the Circle's most famous member – Rudolf Carnap – in a manifesto entitled "The Elimination of Metaphysics Through Logical Analysis of Languages", in which he showed that illusory pseudo-problems of a certain kind of philosophy, for example, Heidigger's "nothing", can in reality be reduced to meaningless words games or nonsensical statements. Or, in Carnap's words, "a music played by musicians without talent".

Among the young people who frequented the Circle was Kurt Gödel, who entered the University of Vienna in 1924 and immediately gravitated into Schlick's orbit. Schlick initiated him by having read Russell's *Introduction to Mathematical Philosophy*. At the university, Gödel also attended Carnap's classes, out of which came Carnap's 1928 *Logical Construction of the World* and the 1934 *Logical Syntax of Language*. As the titles imply, these are two works in which logic was applied to the physical world on one hand, and to human language on the other.

Gödel, who was so curious as a child that he earned the nickname Herr Warum, Mr Why, did not allow himself to be distracted by this kind of secondary problem, and addressed head-on the main questions about the foundations of logic and mathematics raised by Leibniz, Kant, Frege, Russell, Wittgenstein, Hilbert, Poincaré and Brouwer. It is precisely because he chose to focus on these questions definitively, making evident all their weaknesses, that his work is considered to be the most important contribution ever made to mathematical logic.

The first problem that Gödel grappled with was that stated by Hilbert at the International Congress of Mathematicians in Bologna in 1928. He gave the solution the following year, at only 23-years old, in his degree thesis, which contained his first great result: the *incompleteness theorem* for predicative logic, which is analogous to that for propositional logic proven by Post in 1921. To be more precise: analogous to the tautologies are the formulas that are true in all possible worlds which turn out to be exactly the theorems of the predicative system Frege's *Begriffsschrift*, or – if you will – Russell and Whitehead's *Principia*.

Kurt Gödel

Once he had proven the completeness of logic, first propositional and then predicative, the natural thing to do was to extend the results to mathematics, beginning for example by proving that the theorems of the arithmetical system of the *Principia* are precisely the true formulas of arithmetic. Gödel devoted himself to this task in his degree thesis of 1931, but to his surprise he discovered that instead there existed true formulas of arithmetic that were not theorems of the *Principia*.

Even more surprising, however, was that the problem was unsolvable: it was certainly possible to add axioms to the *Principia* in order to render them less incomplete, but there were no possible additions that could have made them complete! For this reason, the title of Gödel's work spoke of "undecidable propositions of *Principia Mathematica* and related systems", because the problem was common to all mathematical systems past, present and future, and not only that constructed by Russell and Whitehead.

As fate would have it, Gödel made the first official announcement of his theorem on 7 September 1930 in Königsberg, on the occasion in honour of Hilbert, who the next day, unawares, pronounced his motto "We must know, and we will know", unaware that by now it was known that not everything can be known.

The idea of Gödel's proof was a variation on the theme of the liar paradox, suitably modified so that it became a theorem. Where Eubulides had considered a sentence ("this sentence is false"), Gödel considered a formula ("this formula cannot be proved"). Naturally, since there is only one truth – or there is only one if there are any at all – Eubulides' sentence is paradoxical but not ambiguous. However, there are many proofs: one for each system of axioms and rules. Gödel's formula is thus ambiguous and must be restated in light of a particular system, for example, that of the *Principia*, and saying, "This formula cannot be proved in the given system".

Eubulides asked if his sentence were true or false, and discovered that neither case was possible. Analogously, Gödel asked if his formula could be proved or disproved, and he too discovered that neither of the two cases is possible, at least if the system proved only truth. Because, in this case, if the formula could be proved it would be true and thus not provable. So it cannot be proved, and thus is true: that is, in the system there are truths that cannot be proved, exactly like in the best criminal trials of the Mafia.

But Gödel's formula cannot only not be proved in the system, but it also cannot be disproved, because not even its negation can be proved: it is in fact false, and the system only proves truth. So, the system includes formulas that can be neither proved nor disproved. Such formulas are examples of those perennially undecidable statements whose existence was intuited by Brouwer. Of, if you prefer, we might say that the principle of the excluded third is not valid for provability, because Gödel's formulas are in fact examples of the third party in a disagreement between two parties, "can be proved" and "cannot be proved".

Naturally, in order for this reasoning to work, it is not sufficient that the system considered does not prove falseness: it must also make it possible to express formulas that say that they cannot be proved in the system. But Gödel discovered that just one small thing was needed to make this possible. When the system has a minimum capacity for expression, its syntax can be reduced to arithmetic in a way that was analogous to that prefigured by Leibniz, that is, assigning simple numbers to simple terms and composite numbers to composite terms.

Leibniz had assigned products to composite terms, without taking into account the fact that in multiplication the factors are lost and it becomes impossible to find them again unequivocally. Gödel got around the problem by taking advantage of

Euclid's theorem, according to which the decomposition of a number into prime factors is unique, and he thus assigned to composite terms products of prime numbers having as exponents the numbers of the components. He was therefore also able to prove, in passing, that Wittgenstein was mistaken in the *Tractatus* when he said that language could not speak about its own logical form, at least if by logical form what is meant is syntactic structure.

In addition to making it possible to reduce its own syntax to arithmetic, the given system must also make it possible to construct formulas that speak about themselves. In natural language the problem doesn't arise, because pronouns such as "I", or adjectives such as "this" immediately make the construction of the sentence self-referential, such as "I lie" or "this sentence is false". In mathematics, the situation is more complicated, but not impossible. For example, any equation in which a variable appears both to the right and the left of the equal sign constitutes a self-referential definition of the solution of the equation. And it is precisely thanks to the fact that numbers are assigned to the formulas that Gödel is able to prove that formulas such as "this formula cannot be proved" can be obtained by solving appropriate equations.

Once we are aware of these circularities, however, it is not difficult to spot them in many other areas. For example, in computer science the arrows in flow diagrams and the command "go to" make it possible to construct a loop in the program. In cybernetics, feedback takes account of the system's homeostasis, that is, its capacity to maintain equilibrium by strengthening internal connections or weakening external agents. In biology, autopoiesis, or self-creation, describes an organism's capacity to reproduce itself. In chemistry, catalytic cycles or loops, in which the product of a reaction is involved in its own synthesis, are responsible for a system's instability, and thus, in the final analysis, for the instability of life. Finally, in physics, the entire universe can be interpreted in terms of a self-excited circuit, one which generates the observer who generates that which is being observed.

It is precisely because Gödel's proof uses instruments that are so pervasive that the theorem of incompleteness became a paradigm of an entire school of thought. As a consequence, it has become one of the few – not to say only – results in mathematics to be referred to in a musical composition, such as Hans Werner Henze's second violin concerto; or in poetry, as in Hans Magnus Enzensberger's "Homage to Gödel"; or in film, as in Fred Schepisi's 1994 "I.Q."; or in science fiction novels, such as Stanislaw Lem's *Golem XVI*, Rudy Rucker's *Software*, William Gibson and Bruce Sterling's *The Difference Engine*, and Samuel Delaney's *The Einstein Intersection*, in addition to many others.

In any case, a theorem that deals with the impossibility of proving theorems is a typical cultural expression of the twentieth century, a century that has seen all kinds of artists describe the limits of expressions of their particular medium through the medium itself. For example, prime example are is Luigi Pirandello's *Six Characters in Search of an Author* in literature, Federico Fellini's film "8 and ½" in cinema, John Cage's "4′33″" in music, and the monochromes by Yves Klein in painting.

Of course, the impossibility of completely describing a sufficiently complex reality had already been largely anticipated. Example are found in *Poetica* by Artistotle in literature, in Kant's *Critique of Pure Reason* in philosophy. Indeed, the incompleteness theorem can be considered as a reformulation and a formalisation of the assumed principle of Kant's *The Transcendental Dialect*: that is, the fact that in order for reason to be complete and make it possible to consider transcendental ideas, it has to be inconsistent and fall into the antimonies of pure reason.

A strengthened version of the incompleteness theorem, proved by John Barkley Rosser in 1936 with an argument that was analogous to that of Gödel, although slightly more complicated due to its being based on the formula, "this formula cannot be proved before its negation", shows that, if a mathematical system with a minimum expressive capacity wants to be consistent and not lapse into contradiction, then it must be incomplete.

On the other hand, as far as Kant is concerned, the incompleteness theorems prove that mathematics cannot be reduced to logic, for which instead a completeness theorem holds. One of the philosophical consequences of Gödel's theorems is thus the definitive proof that the logic dreamed of by Frege and Russell cannot become reality, and that instead Kant and his followers, from Poincaré to Brouwer, were right: arithmetic is not a priori analytic but synthetic. There is no more discussion about this.

At least for us, that is, because neither Russell nor Wittgenstein understood antiphony and even less, the psalm. His whole life long Russell believed that Gödel had proved that arithmetic was inconsistent, while Wittgenstein thought that there was something the matter with the whole business, because you couldn't prove that something couldn't be proved. At which point Gödel was forced to respond that both men were pretending to be stupid, unless they actually were.

As far as Gödel himself was concerned, not a man to slight anyone, he destroyed Hilbert's program on consistency, showing that it too was impossible to realise: no mathematical program that is consistent and has a minimum expressive capacity can prove its own consistency. In fact, the theorem of incompleteness is based on the hypothesis which says that if a system is consistent, then no given formula can be proved: thus, if the hypothesis could be proved, then the thesis (that the formula could be proved) could also be proved. But that formula states precisely that it cannot be proved; therefore the formula itself could be proved, and instead it cannot.

But if a system cannot prove its own consistency, this means that it cannot justify itself, and thus that its own justification lies outside itself: that is, there are no Baron von Münschausens in mathematics (academics, yes, after 1968). In particular, there is no hope of stopping the game of passing the buck that Hilbert had tried to put a stop to when he proposed, at the International Congress of Mathematicians in Paris, proving the consistency of arithmetic or analysis directly and with elementary instruments. In other words, in a single blow Gödel had also solved Hilbert's second problem by proving that it could not be solved.

Of course, these arguments are so subtle that they certainly caused headaches for this author, and probably for the reader as well. Just imagine what happened to the one who worked to discover and prove them, that is, Gödel himself. Gödel, at the age of six, had already shown evidence of some mental problems when, following rheumatic fever that the doctors said he had recovered from completely, he was convinced instead that he had been left with a permanent lesion to his heart.

This was the beginning of the hypochondria and mistrust of doctors that he harboured throughout the rest of his life, and of a mental fragility that led to his being more than once admitted to a psychiatric hospital, starting right from the early years of the 1930s, when the effort he made to concentrate on his first theorems led to a mental collapse. When he recovered, the only thing that one who had already solved Hilbert's second problem could do was solve the first, that is, that of the continuum hypothesis.

Gödel tried, but this time he was only halfway successful: that is, he proved that the continuum hypothesis could not be disproved in the axiomatic system for set theory, which had been developed starting in 1908 by Ernst Zermelo, a student of Hilbert's, and which from that time on had been the usual point of reference for mathematicians working on these things.

To be more precise, Gödel constructed a world of sets that satisfied both the axioms of Zermelo and the continuum hypothesis; that is, one in which there are no infinities whose cardinality is between those of integers and real numbers. This world is inspired by the seventh point of Wittgenstein's *Tractatus*, because this contains only sets which can be spoken about only in the language of sets: in other words, there are sets that must be found in all possible worlds of sets, but nothing else.

If in this minimal world there already were infinities between those of integers and real numbers, then the discussion would have ended there, and the continuum hypothesis would have been disproved. But instead Gödel proved that there weren't any, leaving open two possibilities: either that the hypothesis could be proved – and thus that it was true not only in its own, but in all possible worlds – or that it could be neither proved nor disproved, because it was true in its own world but false in another.

In 1963 Paul Cohen proved that the second possibility was the correct one, constructing various alternatives to the minimal world of Gödel, alternatives in which there were any number of infinities between those of integers and real numbers. The problem that Hilbert considered to be the most difficult one of modern mathematics was thus solved in the same way as the second had been: that is, by discovering that it could not be solved within the usual theory of sets; so much for "*non ignorabimus*".

The possibility that the continuum hypothesis might be undecidable had actually already been suspected by the Norwegian Thorald Skolem in 1922, when he noted an interesting phenomenon: that one of the possible worlds of set theory contained exactly as many elements as the set of integers and no more. Now, a world of sets has to contain a lot of things, among them the real numbers. Real numbers, however, already on their own have a greater infinity than integers. Where is the catch?

At first a new paradox was feared, but then Skolem came to understand that, simply, the real numbers of that world were not the "true" real numbers but only a set whose properties were the same as the set of real numbers. Analogously, the "infinities" of that world were not the "true" ones, and the fact that there appeared to be more "real numbers" than integers only meant that in that world there was no bijection, or one-to-one correspondence between them.

In fact, from within that world, the infinity of infinities whose existence had been proven by Cantor *appear to be different from one another*, but from outside *they are all equal to* the infinity of integers. And since Skolem's world of sets is just as valid as any other, it is quite possible to think that there is actually only one infinity, that of the integers known since antiquity, and that the "superinfinities" introduced by Cantor are fictional. Or better, that we might say they come down to signalling not the *presence* of many objects, but the *absence* of many one-to-one correspondences.

So, once again logic took its anti-metaphysical stand and was able to deconstruct Cantor's theory of infinities, a theory which had even caused discern in the Catholic church. To be more precise, it was no longer necessary to interpret the theory, which asserted that there were many infinities, in an ontological-positivistic way, and it was possible to consider it in an epistemologically negative way: that is, as one of the many products of the limits of mathematical thought, that is, in line with incompleteness proved by Gödel's.

When in 1938 Hitler invaded Austria, Gödel found himself under German rule. To his surprise, the military doctors did not agree with his self diagnosis of a heart condition and he was deemed to be fit for the draft, with the risk that he would have to serve in the trenches. When he finally decided to flee, as many other members of the Vienna Circle had already done, war had already broken out, and in order to reach Princeton he had to cross the Soviet Union by train, the Pacific Ocean by ship, and the United States, again by train.

That journey having completely satisfied any need he might have had for adventure, Gödel never returned to Europe and steadfastly refused any award offered to him by Austria, although not for the reasons that appear obvious: when Gödel arrived in Princeton, he was met by the economist Morgenstern, another emigrant from Austria, who asked him how the situation in Austria was. Gödel's answer was that the coffee was awful.

Having settled in 1939 at Princeton's Institute for Advanced Study, he interpreted the rules in his own way and devoted himself to philosophy. By reading Kant he was stimulated to discovery of one of his most surprising results, which led to his winning the Einstein medal: the possibility of travelling through the past in keeping with the general theory of relativity, which proved that Kant was right in considering time not as a physical reality, but rather as an a priori form of our senses. On the other hand, in Leibniz, who Gödel considered to be an extremely gifted philosopher because "he had gotten it all wrong", he found the inspiration for a mathematical proof of the existence of God. According to his wife, however, one of the things he was most deeply interested in was demonology.

This wife was a divorced dancer, older than Gödel, with whom he had fallen in love when still a student but was able to marry only in 1938 because his parents had opposed the marriage. She must have had a good sense of irony, since once she said to him at a meeting, "Kurtele, if I compare your lecture with the others, there is no comparison". To be sure, her presence contributed to Gödel's emotional stability, and when she was admitted to hospital in the 1970s, his depression and paranoia were given free rein. He got it into his head that someone was trying to poison him, and died in 1978 of "malnutrition caused by personality disturbances".

As shown by the fact that today his name is too often taken in vain, Gödel was a god of logic; it is also true that his name is can be read as "God" and "El", which means God in English and Hebrew. If we want to compare him to some great god of the past, the first who springs to mind is that prince of mathematicians, Gauss: both published very few papers, in keeping with the motto *pauca sed matura,* few but good, and both kept results in the drawer that anyone else would have boasted about. If the most obvious parallel is Aristotle, the more appropriate one is Archimedes: neither of the two created his own discipline, but both changed his chosen discipline forever with his results, thus achieving a depth that is apparently unfathomable.

Hommage À Gödel

| Hommage à Gödel | Homage to Gödel |
Hans Magnus Enzensberger	tr. the poet and Michael Hamburger
Münchhausens Theorem, Pferd, Sumpf und Schopf, ist bezaubernd, aber vergiß nicht: Münchhausen war ein Lügner.	"Pull yourself out of the mire by your own hair": Münchhausen's theorem is charming, but do not forget: the Baron was a great liar.
Gödels Theorem wirkt auf den ersten Blick etwas unscheinbar, doch bedenk: Gödel hat recht.	Gödel's theorem may seem, at first sight, rather nondescript, but please keep in mind: Gödel is right.
"In jedem genügend reichhaltigen System lassen sich Sätze formulieren, die innerhalb des Systems weder beweis- noch widerlegbar sind, es sei denn das System wäre selber inkonsistent."	"In any sufficiently rich system statements are possible which can neither be proved nor refuted within the system, unless the system itself is inconsistent."
Du kannst deine eigene Sprache in deiner eigenen Sprache beschreiben: aber nicht ganz. Du kannst dein eignes Gehirn mit deinem eignen Gehirn erforschen: aber nicht ganz. Usw.	You can describe your own language in your own language: but not quite. You can investigate your own brain by means of your own brain: but not quite. Etc.
Um sich zu rechtfertigen muß jedes denkbare System sich transzendieren, d.h. zerstören.	In order to be vindicated any conceivable system must transcend, and that means, destroy itself.
"Genügend reichhaltig" oder nicht: Widerspruchsfreiheit ist eine Mangelerscheinung	"Sufficiently rich" or not: Freedom from contradiction is either a deficiency symptom,

(*continued*)

C. Bartocci et al. (eds.), *Mathematical Lives*,
DOI 10.1007/978-3-642-13606-1_11, © Springer-Verlag Berlin Heidelberg 2011

Hommage à Gödel	Homage to Gödel
Hans Magnus Enzensberger	tr. the poet and Michael Hamburger
oder ein Widerspruch.	or it amounts to a contradiction.
(Gewißheit = Inkonsistenz.)	(Certainty = Inconsistency.)
Jeder denkbare Reiter,	Any conceivable horseman,
also auch Münchhausen,	including Münchhausen,
also auch du bist ein Subsystem	including yourself, is a subsystem
eines genügend reichhaltigen Sumpfes.	of a sufficiently rich mire.
Und ein Subsystem dieses Subsystems	And a subsystem of this subsystem
ist der eigene Schopf,	is your own hair,
dieses Hebezeug	favourite tackle
für Reformisten und Lügner.	of reformists and liars.
In jedem genügend reichhaltigen System,	In any sufficiently rich system
also auch in diesem Sumpf hier,	including the present mire
lassen sich Sätze formulieren,	statements are possible
die innerhalb des Systems	which can neither be proved
weder beweis- noch widerlegbar sind.	nor refuted within the system.
Diese Sätze nimm in die Hand	These are the statements
und zieh!	to grasp, and pull!

English translation by Hans Magnus Enzensberger and Michael Hamburger. From: *Selected Poems*, Sheep Meadow Press, 1999.
© Suhrkamp Verlag; Trans. © the poet & Michael Hamburger.
Reprinted with kind permission of Suhrkamp Verlag.

(Michael Hamburger died on 7 June 2007 at his home in Suffolk. Hans Enzensberger is still living.)

Robert Musil

The Audacity of Intelligence

Claudio Bartocci

The Country that Went to Ruin Because of a Linguistic Gap

"Kakania was, after all, a country for geniuses; which is probably what brought it to its ruin".[1] In the pages of *The Man without Qualities*, Kakania (a neologism coined by Musil from the abbreviation "k.k." for *kaiserlich-königlich*, imperial and royal) is the ironic and scatological name used to indicate the waning Austro-Hungarian monarchy.

The new century, the twentieth, by many too hastily hailed as the era of the definitive triumph of Western civilization and technological progress, appeared to have been left in ruins almost as soon as it had started by the appalling bloodbath of the First World War. Karl Kraus entitled his hypertrophic satirical–apocalyptical drama *The Last Days of Mankind*, while Stefan Zweig evoked the twilight of the old Austria in *The World of Yesterday*. In spite of this, from many points of view the Great War did not represent a clean break or essential discontinuity in Austrian culture as much as a difficult moment of transition, a painful passage that left its basic nature unchanged. This is not merely the persistence (at least in the context of literature, but not only) of a "Habsburgian myth" as a fantastic and poetical transfiguration of the era of Franz Josef,[2] but of the perpetuation – in an intellectual class that was not decimated by the war as was the case, for example, in France – of ideas, concepts, and visions of the world: not a single *Weltschauung*, but a variegated complex of *Weltschauungen*, in which it is nevertheless possible to identify some common themes.

---

[1] R. Musil, *The Man without Qualities,* transl. by Sophie Wilkins and Burton Pike, 2 vols., Vintage Books, New York, 1996, vol. I, p. 31. From this point on the work will be cited with the abbreviation *MWQ* followed by the volume and page number (the second volume comprises the section *From The Posthumous papers*, translated by Burton Pike).

[2] C. Magris, *Il mito absburgico nella letteratura austriaca moderna* (rev. ed), Einaudi, Torino, 1988.

C. Bartocci et al. (eds.), *Mathematical Lives*,
DOI 10.1007/978-3-642-13606-1_12, © Springer-Verlag Berlin Heidelberg 2011

Robert Musil

The leitmotifs that characterize the disordered polyphony of the culture of the German language in the first 30 years of the twentieth century can all be traced back (with some in evitable generalizations) to the problem of the identity of the subject and – in strict connection to that – to limits of language, that is, the ability to express the world through words and formulas, of the basis of the discourses that are uttered to describe things.[3] An evident red line connects Hugo von Hoffmannsthal's *The Lord Chandos Letter* ("...the abstract terms of which the tongue must avail itself as a matter of course in order to voice a judgment – these terms crumbled in my mouth like mouldy fungi"[4]) to Wittgenstein's *Tractatus* ("The limits of my language mean

---

[3]The bibliography on this topic is enormous. We will cite only as an example A. Janik and S. Toulmin, *Wittgenstein's Vienna*, Simon and Schuster, New York, 1973, J. Le Rider, *Modernity and Crises of Identity: Culture and Society in Fin-de-siècle Vienna*, transl. by R. Morris, Continuum, New York, 1993 and Carl E. Schorske, *Fin-de-siècle Vienna : politics and culture*, Vintage Books, New York, 1980.

[4]H. von Hofmannsthal, "The Letter of Lord Chandos", in *Selected Prose*, trans. by Mary Hollinger and Tania and James Stern, Routledge and Kegan Paul, London 1952, pp. 133–134. In a letter to Edgar Karg von Bebenburg, Hofmannsthal wrote, "Words are not of this world, they are a world of their own, a world completely independent, like the world of sounds" (H. von Hofmannsthal, *Le parole non sono di questo mondo,* ed. by M. Rispoli, Quodlibet, Macerata, 2004, cf. endnote 40,

the limits of my world"[5]), by way of Fritz Mauthner's[6] *Sprachkritik* and the dazzling aphorisms of Karl Kraus: "once a word has entered into relationship with the world, the end is endless".[7] Kurt Gödel observed, "The more I think about language, the more it amazes me that people ever understand each other".[8]

These motifs – at the root of which are found the reflections of such diverse thinkers as Frege, Mach and Nietzsche – permeate all of Musil's works – from *The Confusions of Young Törless*, to the *Posthumous Papers of a Living Author*, to *The Man without Qualities*. In the pages of this labyrinthine novel, this "work-world",[9] Ulrich – the man who is without qualities because he possesses a sense of the possible even in his own regard – finds himself having to deal with the elusive evanescence of words: "Words leap like monkeys from tree to tree, but in that dark place where a man has his roots he is deprived of their kind mediation" (*MWQ*, vol. I, p. 164). Musil's irony, using the allegory of the linguistic Babel that reigned in Franz Josef's empire, even identifies the main reason for its dissolution in the checkerboard of languages:

> Since the world began, no creature has as yet died of a language defect, and yet the Austrian and Hungarian Austro-Hungarian Dual Monarchy can nevertheless be said to have perished from its inexpressibility (*MWQ*, vol. I, p. 491).

Precision and Soul

For 37 uninterrupted years, from 1899 until his death in 1936, Karl Kraus never ceased to shoot arrows at all the *idées reçues* of Austrian society and culture from the pages of *Die Fackel* (*The Torch*), a small brick-red magazine in which he was the sole author starting in 1911. 1899 was an important year, in which were inaugurated the Karlsplatz station of the underground designed by the architect Otto Wagner, the publication of Freud's *Die Traumdeutung* (*Interpretation of Dreams*) and Hilbert's *Grundlagen der Geometrie* (*Foundations of Geometry*). 1936 – when Nazism celebrated itself at the Olympics in Berlin and grim clouds

---

pp. 99–100); it seems that these words would echo some observations contained in Novalis's *Monologue*.

[5]L. Wittgenstein, *Tractatus Logico-Philosophicus* 5.6.

[6]On Mauthner see A. Janik and S. Toulmin, *Wittgenstein's Vienna*, op. cit, Chap. 5. We also note the polemic remark of Wittgenstein: "All philosophy is a 'critique of language' (though not in Mauthner's sense). It was Russell who performed the service of showing that the apparent logical form of a proposition need not be its real one" (*Tractatus Logico-Philosophicus* 4.0031).

[7]K. Kraus, *Detti e contraddetti* [*Sprüche und Widersprüche*], Adelphi, Milano, 1987, p. 256 (cf. *Dicta and Contradicta*, transl. by J. McVity, University Of Illinois Press, 2001).

[8]Cited in R. Goldstein, *Incompleteness: The Proof and Paradox of Kurt Gödel*, W.W. Norton, New York, 2005.

[9]See F. Moretti, *Modern Epic: The World-System from Goethe to García Marquez*, transl. by Q. Hoare, Verso, 1996, especially Chap. 7, Sect. 5.

gathered over Europe's future – was the year in which the philosopher Moritz Schlick, the founder of the Vienna Circle, was assassinated on the steps of the university by a student who was tied ideologically to the Austro-Fascist movement.

It was in this long and contradictory period that Vienna came to the fore as one of the most vital driving forces of European culture,[10] which is even more surprising given that it was a city that was generally provincial, first the centre of a rapidly declining empire and then the capital of a nation without political ambitions. The cultural scene was dominated by figures who left their mark on modernity: musicians such as Gustav Mahler, Arnold Schönberg, Anton Webern, Alban Berg, Richard Strauss; architects such as Otto Wagner and Adolf Loos; writers such as Hugo von Hofmannstahl, Joseph Roth, Kraus, Arthur Schnitzler, Heimito von Doderer, Hermann Broch, Franz Werfel; artists such as Gustav Klimt, Oskar Kokoschka, Egon Schiele, and then Ernst Mach and Ludwig Boltzmann, the mathematician Hans Hahn, Schlick, Freud, Wittgenstein, Rudolf Carnap, Otto Neurath.

The tables of famous cafés (*Café Griensteidl, Café Central, the Herrenhof*) or the meeting places of the innumerable circles, dinner clubs and cultural associations were home to interwoven discussions, often centred around opposite points of views. The Viennese intellectual milieu was anything but monolithic; it was like a crucible in which material extracted from various cultures was melded, a mosaic, a chorus of voices that were sometimes dissonant. In Kraus's words, Vienna was "the research laboratory for the destruction of the world". Mystics and neo-positivists, symbolists and expressionists, "classical physicists" and creators of new ideas regarding the quantum, supporters and denigrators of psychoanalysis. In this kind of cultural climate, literature and science don't necessarily go hand in hand, as is sometimes claimed: they are opposing and often quite distant universes. To be sure, Mach – with his "analysis of sensations", with his concept of "the unsavable I" (*das unrettabare Ich*) – had a profound influence on writers like Hofmannstahl and Bahr. To be sure, Schnitzler was trained in medicine; Leo Perutz flanked his literary work with his profession as a mathematician in the insurance business for many years; Elias Canetti had a degree in chemistry; Broch, abandoning at 40 the management of his father's textile firm, dedicated himself to serious study of philosophy and mathematics (the leading character of his novel *Die Unbekannte Größe* (*The Unknown Quantity*) is a mathematician). But only Musil, we believe, attempts with lucid awareness to fill the gap between *Dichtung* and *Erkenntnis*,[11] to reconcile the disagreement between "precision and soul". The fundamental tension between literature and science could be overcome and reshaped as a model for novel writing, freed from the narrative conventions of the nineteenth century, if one only knew how to describe the facts, not as they are, but as they *could* be, knew how to capture

---

[10]In spite of this, as E. Gombrich writes, "The thesis that most of the intellectual life of this century [the 1900s] was invented in Vienna is, of course, not worth discussing" (*The Visual Arts in Vienna circa 1900*, "Occasions", vol. 1, The Austrian Cultural Institute, London, 1997).

[11]Cf. L. Dahan-Gaida, *Musil. Savoir et fiction*, Presses Universitaires de Vincennes, Saint-Denis, 1994, p. 17.

"the mystery of what happens".[12] The ideal archetype of this new form of novel – which would become that of *Man without Qualities* – is shown in the following passage:

> The goal of scientific thought is the unambiguous expression and correlation of facts. It is most admirable where it permits one to travel nakedly through its splendid hardness. Essayistic thought can give no contrast to this but should rather be a continuation, authorized to go where scientific thoroughness finds no foundation that will hold firm with the strength essential for its application. . .[13]

Too Intelligent To Be a Poet

In 1901 Musil earned a degree in mechanical engineering at the Brno Polytechnic.[14] His basic scientific training – mathematical analysis, rational mechanics, physics – was probably not very different from that given to Einstein, more or less in the same years, at the Zurich Polytechnic, or to Wittgenstein, some years later, at the Berlin Technisches Hochschule. After a brief, unsatisfying stint as an unpaid assistant at the Technical University of Stuttgart (to while away the time he began writing his novel *The Confusions of Young Törless*), in 1903 he enrolled at Humboldt University in Berlin, where he attended courses in philosophy and psychology under the guidance of Carl Stumpf. Stumpf (1848–1936), who was one of the greatest experts in experimental psychology, and author of works on spatial representation and the physiological and psychological aspects of musical phenomena (his *Tonpsychologie* is fundamental), also undertook important research regarding the structuring of cognitive processes, in close relation to the investigations carried out by Alexius von Meinong and by a group of other scholars including Christian von Ehrenfels, in which it is possible to find traces of the origin for many motifs that would become part of *Gestaltpsycologie*, the "psychology of shape".

The Berlin years were crucial for Musil. In 1905 he completed *Törless*, which was published the following year to good critical acclaim; the same year he also met Martha Marcovaldi, née Heimann (1874–1949), mother of two children from a previous marriage, who became his wife in 1911. Even though the academic world had left him unsatisfied, Musil left off the literary activities that he had planned in order to finish his studies: in 1907 he designed a refined model of a chromatograph (a device used to study the laws of perception of colours) and in 1908 he received

---

[12]Musil in a 1926 interview, quoted in C. Magris, *Il mito absburgico*, op. cit., p. 303.

[13]R. Musil, *Saggistica* (1913), in *Saggi e Lettere,* ed. and with an introduction by B. Cetti Marinoni, 2 vols., Einaudi, Torino 1995, vol. 1, pp. 193–194.

[14]The biographical information regarding Musil is mainly drawn from the *Cronologia della vita e delle opere* in R. Musil, *Diari 1899–1941,* edition by A. Frisé, ed. and with an introduction by E. De Angelis, 2 vols. Einaudi, Torino, 1980, vol. 1, pp. XLI-LVI, and from Wilfried Berghahn, *Robert Musil*, Rowohlt, Reinbek, 2004. The most extensive biography is Karl Corino, *Robert Musil. Eine Biographie*, Rowohlt, Reinbek, 2003.

his doctorate in philosophy with a thesis entitled *Beitrag zur Beurteilung der Lehren Machs* (*On Mach's Theories*). In the meantime, a falling out had occurred with Stumpf, who – perhaps in resentment of the timing of the publication of *Törless* – judged Musil's thesis to be insufficiently critical of Mach and out of line with Stumpf's own rigidly dualistic psychology.[15] In December 1908 Musil was invited by von Meinong, a professor of philosophy in Graz, to become his assistant. Anguishing over the decision, Musil finally declined, deciding instead to pursue a career as a writer: "my love for the literary arts is no less than that for science".[16] In 1910, Musil, by now almost 30, settled in Vienna, accepting a position as assistant librarian at the Polytechnic.

Musil's store of knowledge of philosophy and science went beyond what he had learned during his time at university, which was undoubtedly already even broader in scope than that of the majority of his fellow writers. His *Diaries*, as well as his letters, lectures, essays and *The Man without Qualities*, testify to the wide scope of his interests and his insatiable curiosity. As far as philosophy is concerned, he read Plotinus and the German mystics (don't forget that in the negative theology of Meister Eckhart the supreme being is said to be *ohne Eigenschaften,* "without qualities"), he studied Boltzmann, Husserl and Cassirer, he pored over Aristotle, Leibniz and Nietzsche. He was surely up to date on the studies carried out by the Vienna Circle, although he didn't always share the positions they took. For example, in 1920 in his diary he disdained Otto Neurath ("an academic brawler"[17]), and in 1937 he was harshly critical of the ingenuity of "physicalism" applied to psychology by one of Schlick's students ("how much more precisely things were carried out in Stumpf's school",[18] he noted). Greater affinity is likely to be found with Rudolf Carnap, for example, with his ideas regarding the process through which the knowing subject represents the external world beginning with elementary information. In a letter dated 29 October 1935 Musil wrote: "Among all the books that I have read this year, the one that made the 'biggest impression' was undoubtedly *The Logical Syntax of Language* by Rudolf Carnap".[19] In this book, published in 1934, Carnap formulated his principle of tolerance (also known as the principle of conventionality of language forms), which states that there exist a multiplicity of logical structures or frameworks that can be used to express the various discourses of the empirical sciences, and that it is possible to choose conventions freely among them. Statements of the principle of tolerance that are substantially equivalent (and certainly independent because of the differences in dates) are scattered throughout *Man without Qualities*, especially in the second volume: "What happens doesn't

---

[15]Cf. M. Montinari, "Introduction" to R. Musil, *Sulle theorie di Mach*, Adelphi, Milano, 1981, p. ix.

[16]*Saggi e lettere*, op. cit., vol. II, p. 537.

[17]*Diari, cit.,* vol. I, p. 653.

[18]*Saggi e lettere*, op. cit., vol. II, p. 1369 ('Tagebuch 33', ca. 1937). In 1934 he noted, "Unity of the sciences, where was this dogma come from?" (*Ibid.,* vol. II, p. 1312).

[19]*Ibid.,* vol. II, p. 804.

really matter?" Ulrich tells Agatha, "Waht counts is the system of ideas by which we understand it, and the way it fits into our personal outlook" (*MWQ*, vol. II, p. 743). Both Carnap and Musil claimed that "It is not our business to set up prohibitions' (*Wir wollen nicht Verbote Aufstellen*), but to arrive at conventions".[20]

Profound and determinant were the influences exerted by *Gestaltpsychologie* on the works of Musil, especially as it has been formulated in the work of Wolfgang Köhler (another of Stumpf's students), Kurt Koffka and Max Wertheimer, the three most authoritative scholars of the Berlin school as opposed to the Graz school, whose members were students of Meinong. In 1920 Musil read Köhler's book *Die physischen Gestalten in Ruhe un im Stationären Zustand* (*Physical Gestalt in Rest and Stationary State*), defining it as "extraordinary",[21] in which can be detected an echo of the teaching of a theoretical physicist such as Max Planck. In the long essay of 1921 entitled *Wege zur Kunstbetrachtung* (*How to Approach Art*), Musil gave the work with the same title by Johannes von Allesch (Musil's schoolmate in Berlin and one of his closest friends as a youth) a very positive review, recognising "its incomparable merit . . . of having founded for the first time a method that is flexible yet resistant for aesthetic evaluation".[22] As Claudio Magris writes, ". . . from Gestalt theory Musil takes the idea that everything is given as wholes before parts, as totalities that are perceived individually, as an immediate relation on the part of all and not as a sum of parts".[23]

As far as the exact sciences are concerned, Musil kept himself up to date. "Robert is presently very busy with Einstein's theories, but he is looking for another way", wrote his wife Martha to her daughter Annina Marcovaldi on 17 May 1923.[24] But even in his essay "The Mathematical Man", written 10 years earlier, there is explicit mention of the new ideas of relativity, cited as an example of the "confidence and pride in the devilish riskiness of his intellect" with which the scientist addresses the scandals of reason:

> I could adduce still other examples, for instance when mathematical physicists were suddenly wildly bent on denying the existence of space and time. But they did not do this in a dreamy haze, the way philosophers sometimes do (which everyone then immediately excuses by saying: Look at their profession), but with reasons that rose up before us quite suddenly as palpably as an automobile, and became terribly credible.[25]

In May 1923 Musil sent a "note on quantum theory" (unfortunately not preserved in the *Nachlass*) to Annina: "it certainly won't serve to explain it to you, and

[20]R. Carnap, *The Logical Syntax of Language*, transl. by Ametha Smeaton, Open Court Publishing, Chicago, 2002, p. 51.

[21]*Saggi e lettere*, op. cit., vol. II, p. 604.

[22]Ibid., vol. II, p. 268.

[23]C. Magris, *L'anello di Clarisse. Grande stile e nichilismo nella letteratura moderna*, Einaudi, Torino, 1999, p. 233.

[24]*Saggi e lettere*, op. cit., vol. II, p. 635.

[25]R. Musil, *Precision and Soul: Essays and Addresses*, ed. and transl. by Burton Pike and David S. Luft, University of Chicago Press, Chicago, 1995, p. 39.

at the moment not even I know more about quanta then is written here, but perhaps, when you will have read it, it will give you a first sense of familiarity with the subject".[26]

With regard to mathematics – described figuratively as "the bold luxury of pure reason"[27] – Musil was undoubtedly aware of the new studies in set theory and the research in the foundations of mathematics, and his awareness was not superficial:

> And suddenly, after everything had been brought into the most beautiful kind of existence, the mathematicians – the ones who brood entirely within themselves – came upon something wrong in the fundamentals of the whole thing that absolutely could not be put right. They actually looked all the way to the bottom and found that the whole building was standing in mid-air. But the machines worked! We must assume from this that our existence is a pale ghost; we live it, but actually only on the basis of an error without which it could not have arisen. Today there is no other possibility of having such fantastic, visionary feelings as mathematicians do.[28]

Musil's reflections on notions of chance and probability[29] are very important and essential to understanding that concept of the "sense of the possible" that is central to *Man without Qualities*. Beginning with Boltzmann's kinetic theory of gas, and taking a critical look at the works of Neurath, Carnap and von Mises (whose circle he frequented during his stay in Berlin in the years 1931–1933), Musil arrives at a statistical concept of the notion of probability based on the "law of large numbers"[30] and in clear contrast with the subjective concept (epistemic) as well as the objective concept (ontological).[31]

The scientific concepts, and in particular those drawn from mathematics and physics, don't figure merely as accessories in Musil's work, but are fundamental, providing a program for the process of literary creation itself:

> But all intellectual daring today lies in the natural sciences. We shall not learn from Goethe, Hebbel or Hölderlin, but from Mach, Lorentz, Einstein, Minkowski, from Couturat, Russell, Peano...
>
> And from within the program of this art the program of an individual work of art might be this:
>
> Mathematical daring, dissolving souls into their elements and unlimited permutation of these elements; here everything is related to everything else and can be built up from these elements.[32]

---

[26]*Saggi e lettere*, op. cit., vol. II, p. 634 (the note was attached to the same letter from Martha cited in note 24).

[27]R. Musil, "The Mathematical Man", in *Precision and Soul: Essays and Addresses*, op. cit., p. 41.

[28]R. Musil, "The Mathematical Man", in *Precision and Soul: Essays and Addresses*, op. cit., p. 42.

[29]An essential treatment of these notions is found in the essay by J. Bouveresse, *L'homme probable. Robert Musil, le hasard, la moyenne et l'escargot de l'histoire*, Éditions de l'Éclat, Combas, 1993.

[30]In the play *The Fanatics* [*Die Schwärmer*, 1921], Stader sets out to found 'the scientific organisation of the universe' based on the law of large numbers.

[31]Cf. "Tagebuch 10" in *Diari*, op. cit, pp. 694–708.

[32]R. Musil, "Profile of a Program", in *Precision and Soul: Essays and Addresses*, op. cit., p. 13.

It is however also true that Musil often did not take kindly to being considered an "essayist" imbued with scientific ideas, or still worse, a philosopher; he shielded himself, defending the specifically poetic nature of his work. In a letter thanking Einstein written in 1941, for example, he defined himself with Nietzsche's words, *Nur Narr! Nur Dichter!* (only a jester, only a poet!).[33] However, this seems to imply that this and similar declarations conceal a truth that is the opposite of their literal meaning, and perhaps they should be taken as a kind of ironic (and vaguely self-complacent) reply to the judgment of the German academy of poets, which had rejected his candidacy for membership on grounds that he was "too intelligent to be a poet".[34]

Integrals To Grow Lean Again

Already in *Törless*, mathematics is the main tool for critical investigation, and at the same time, a metaphor for a knowledge that is *other*, almost a bridge without any visible means of support suspended over an abyss, like in the famous passage on the strange "stuff about imaginary numbers":

> Just think about it for a moment: in that kind of calculation you have very solid figures at the beginning which can represent metres or weights or other measures. And there are real numbers at the end of the calculation as well. But they're connected to one another by something that doesn't exist. Isn't it like a bridge consisting only of the first and last pillars, and yet you walk over it as securely as though it was all there? . . . For me there's something dizzying about a calculation like that, but the really uncanny thing about it is the strength that exists in such a calculation, holding you so firmly that you land safely in the end.[35]

This is the "audacity" that is proper to mathematics, and it includes activities which, although at first sight don't appear to be of any use at all, are actually some of the "most entertaining and intense adventures of human existence".[36] Thus, the work of the mathematician – whose specific nature Musil is able to capture perhaps better than any other writer – is rather like that of an acrobat:

> Ulrich, meanwhile, was at home, sitting at his desk, working. He had got out the research paper that he had interrupted in the middle weeks ago [. . .] He had drawn the curtains and was working in the subdued light like an acrobat in a dimly lit circus arena rehearsing dangerous new somersaults for a panel of experts before the public has been let in. The precision, vigour, and sureness of this mode of thinking, which has no equal anywhere in life, filled him with something like melancholy (*MWQ*, vol. I, p. 115).

---

[33]*Saggi e lettere*, op. cit., vol. II, p. 970.

[34]A judgment similar to this was expressed by Benjamin, who, in a letter to Scholem, defined Musil as "more intelligent than necessary" (cited in E. De Angelis, *Robert Musil. Biografia e profilo critico*, Einaudi, Torino, 1982, p. 45, n. 1).

[35]R. Musil, *The Confusions of Young Törless*, transl. by Shaun Whiteside, Penguin Books, New York, 2001, p. 82.

[36]R. Musil, "The Mathematical Man", in *Precision and Soul: Essays and Addresses*, op. cit., p. 41.

But mathematics does not only reflect the fact that "thinking itself is a vast an undependable affair",[37] it also constitutes an antidote to the sterile nihilsim of thinking, a healthy way of protecting oneself against the spreading *Kitsch* of bad literature:

> We beat the drum for feeling against intellect and forget that without intellect – apart from exceptional cases – feeling is as dense as a blockhead [dick wie ein Mops]. In this way we have ruined our imaginative literature to such an extent that, whenever one reads two German novels in a row, one must solve an integral equation to grow lean again.[38]

In the formalist conception of mathematics, theorised by David Hilbert in order to find a way out of the so-called "foundational crisis", the binomial of syntax/semantics of ancient tradition was demolished, and the central problems became those relative to the coherence and completeness of the axiomatic system: the axioms establish a set of relationships between entities that are abstract, primitive, indefinite, genuine ontological unknowns that correspond to neither things or facts. As Hilbert himself wrote, "[I]t is surely obvious that every theory is only a scaffolding or schema of concepts together with their necessary relations to one another, and that the basic elements can be thought of in any way one likes. If in speaking of my points I think of some system of things, e.g. the system: love, law, chimney-sweep … and then assume all my axioms as relations between these things, then my propositions, e.g. Pythagoras' theorem, are also valid for these things".[39] Because of the arbitrariness intrinsic to mathematics (it was no coincidence that Valéry observed in his *Cahiers* that "les mathématiques sont le modèle de l'arbitraire", mathematics is the model of arbitrariness) it is a discipline that is indissolubly linked to that "sense of the possible" that is the linchpin of Musil's novels: it is in fact "a construction of *possible* orders, a priori to any thought considerations regarding applications and any 'natural' foundations".[40] The freedom of mathematics to create its own theories, limited only by the obligation to respect the coherence of the axioms, suggests that this is the case in real life as well, given that the premises have never been necessary but are rather accidental:

> It might even be fair to say that they were tricked, since nowhere is a sufficient reason to be found why everything should have turned out the way it did; it could just as well have turned out differently (*MWQ*, vol. I, pp. 136–137).

The principle of "sufficient reason" that Ulrich states in his conversation with director Leo Fischel (*MWQ*, vol. I, p. 140) is thus not only a sneer at Leibniz, but

---

[37]*Ibid.*, p. 40.

[38]*Ibid.*, p. 42.

[39]Letter from Hilbert to Frege dated 29 December 1899, as excerpted by Frege in G. Frege, *Philosophical and Mathematical Correspondence*, ed. by G. Gabriel, et al., Basil Blackwell Publisher, Oxford, 1980, p. 40.

[40]M. Cacciari, "L'uomo senza qualità" in *Il romanzo. Volume quinto. Lezioni,* ed. by F. Moretti, Einaudi, Torino, 2003, p. 503.

also an indispensable philosophical instrument for measuring himself against a reality which conceals "a senseless craving for unreality" (*MWQ*, vol. 1, p. 311).

Ulrich cultivates two rather different branches of mathematics. He is described to us both as a mathematical physicist interested in fluid mechanics (cf. *MWQ*, vol. I, p. 115 and vol. II, p. 746: "his eyes immediately alighted on the equations in hydrodynamics where he had stopped"), and as "one of those mathematicians called logicians, for whom nothing was ever 'correct' and who were working out new theoretical principles" (*MWQ*, vol. II, p. 939).[41] It might seem curious that such disparate interests co-exist in the same person, but on careful reflection it can be seen that both of these areas of mathematics provide clues to something specific regarding the problem mentioned earlier, of the impossibility of expressing the world through language. The aim of mathematical physics – as Heinrich Hertz points out in the introduction to his *Principles of Mechanics* – is to construct "models" (*Bilder* or *Darstellungen*) rather than "representations" (*Vorstellungen*) of phenomena, thus it constitutes the necessary technical premise for Ulrich's "distinctly statistical" nature, which is reflected, for example, in what he says to Gerda when he tries to seduce her:

> Suppose the moral sphere works more or less like the physical, as suggested by the kinetic theory of gases: everything whirling around at random, each element doing what it will, but as soon you work rationally what is least likely to result from all this, that's precisely the result you get! Such correspondences, strange as they are, do exist. So suppose we also assume that there is a certain number of ideas circulating in our day, resulting in some average value that keeps shifting, very slowly and automatically – it's what we call progress, or the historical situation (*MWQ*, vol. I, p. 535).

In addition to this, mathematical physical reasoning helps in becoming familiar with those "mathematical problems that do not admit of a general solution but do allow for particular solutions, which one could combine to come nearer to a general solution" (*MWQ*, vol. I, p. 388). When "the problem of human life" is also addressed in these terms, Ulrich – who, like his sister Agathe, is "hal-integrated with himself, a person of 'piecemeal passions'" (*MWQ*, vol. II, p. 766) – is able to recompose the jumbled tesserae of the mosaic of reality, overcoming the intrinsic limits of language, and is able to "look at the world with the eyes of the world", so that these are not "meaningless single things, individual elements, that are as sadly separated from one another as the stars in the night".[42]

On the other hand, "logistics" – emblematic of the "spiritual daring" – opens the way to the "utopia of precise life", teaching how, by means of the inflexible rigour of reason, "to be demanding with ourselves" and shows, through the indissoluble unity of a "sense of reality" and "sense of possibility", that the world cannot be taken literally *sic et simpliciter*:

---

[41]However, Musil adds that Ulrich "he was not entirely satisfied with the logic of the logicians either. Had he continued his work, he would have gone back to Aristotle; he had his own views of all that" (*MWQ*, vol. II, p. 939).

[42]R. Musil, "Tonka", in *Tonka and Other Stories*, transl. by E. Wilkins and E. Kaiser, Picador, London, 1988, p. 298.

God does not really mean the world literally; it is a metaphor, an analogy, a figure of speech that He has to resort to for some reason or other, and it never satisfies Him, of course. We are not supposed to take Him at his word, it is we ourselves who must come up with the answer for the riddle He sets us (*MWQ*, vol. I, p. 388).

It's a dilemma: "... every word demands to be taken literally, otherwise it decays into a lie: but one can't take words literally, or the world would turn into a madhouse!" (*MWQ*, vol. II, p. 813). Not only that, but it's a dangerous dilemma, as shown by the character of Moosbrugger, a mad criminal who calls a squirrel a "fox" or a "hare", in the literal sense of the animal's name in the German dialect, and that of Clarisse, who is destined to go mad, because of the "wretched streak of genius" in her, "the secret cavern where something calamitous was tearing at chains that might one day give away" (*MWQ*, vol. I, p. 155). In the dense network of dis-onthologized relationships that make up the world, only metaphor – "the gliding logic of the soul" (*MWQ*, vol. I, p. 647) – appears capable of providing a solution, but this is a fallacy that fails to untie the knotty relationships between "precision and soul":

A metaphor holds a truth and an untruth, felt as inextricably bound up with each other. If one takes it as it is and gives it some sensual form, in the shape of reality, one gets dreams and art; but between these two and real, full-scale life there is a glass partition. If one analyzes it for its rational content and separates the unverifiable from the verifiable, one gets truth and knowledge but kills the feeling (*MWQ*, vol. I, pp. 634–635).

Perhaps there is no solution, "die Welt, wie sie ist, allenthalben gebrochen erscheint durch eine Welt, wie sie sein könnte".[43] It is useless to wield the blade of irony; useless to string together reasonings:

[Ulrich] had no illusions about the value of his philosophical experimentation; even if he observed the strictest logical consistency in linking thought to thought, the effect was still one of piling one ladder upon another, so that the topmost rungs teetered far above the level of natural life. He contemplated this with revulsion (*MWQ*, vol. I, pp. 648–649).

Musil's novel remains open, not unfinished but in-completed; but it is also true that the great narrative machine of *The Man without Qualities*, like the mathematical method described by Wittgenstein, "is not a vehicle for getting anywhere".[44] Ulrich and Agathe are the "unseparated and not united" (*MWQ*, vol. II, p. 1391); their "journey to paradise" is an immobile permanence, in silence, suspended in the atemporality of the "other state" (*der andere Zustand*). "The truth is not a crystal that can be slipped into one's pocket, but an endless current into which one falls headlong" (*MWQ*, vol. I, p. 582).

---

[43]R. Musil, *Der Mann ohne Eigenschaften*, ed. by Adolf Frisé, 2 vols., Rowohlt, Reibek, 1992, vol. 2, p. 1337: "the world as it appears is everywhere shattered by the world as it could be".

[44]Friedrich Waismann, *Ludwig Wittgenstein and the Vienna Circle*, Brian McGuinness, ed., Basil Blackwell Publisher, Oxford, 2003, p. 33.

The Life, Death and Miracles of Alan Mathison Turing

Settimo Termini

The life of Alan Turing is described in many biographies. The best and most encyclopaedic of these is that of Andrew Hodges; quite pleasant is the agile volume by Gianni Rigamonti, *Turing, il genio e lo scandalo* (Flaccovio editore, Palermo, 1991). Both of these also make mention of his tragic end, which certainly casts a shadow on the mores English society at the time; but of course, who knows how other societies might have behaved?

And the miracles? Well, yes, he worked those too, or at least – if we want to give credit to Kurt Gödel – he worked hand in glove with the other logicians of the first half of the twentieth century, and as far as we know even a single miracle is enough to merit beautification.

In the introduction to one of his books on the theory of computer science, Martin Davis wrote that:

> It is truly remarkable (Gödel...speaks of a kind of miracle) that it has proved possible to give a precise mathematical characterization of the class of processes that can be carried out by purely mechanical means. It is in fact the possibility of such a characterization that underlies the ubiquitous applicability of digital computers.

Davis – a great American logician and computer scientist to whom are owing many important results, making it possible for Yuri Matiyasevich to take the final step towards solving Hilbert's tenth problem – was a student of Emil Post, another great mathematician and logician who (between one depression and the next) contributed to the development of the admirable symphony that is the theory of computability. Post, like Turing, had some very unhappy moments, but perhaps we just can't understand how the creation of such high constructions can make any kind of difficult moment fade into the background.

But what part does Turing play in any of this? He plays a part because he was one of the creators and founders of this theory, along with Alonzo Church, Stephen Cole Kleene and Gödel. Of all of these, Turing was the one who, from the very beginning, most firmly believed in the generality of this theory and its revolutionary

C. Bartocci et al. (eds.), *Mathematical Lives*,
DOI 10.1007/978-3-642-13606-1_13, © Springer-Verlag Berlin Heidelberg 2011

importance. By analysing the way in which a human being proceeds when he has to perform an arbitrary computation, he extracted some basic, essential elements and, idealising them, created an abstract model of a machine, called the Turing machine.

Alan M. Turing

Not content with this, he also stated a thesis, known as the Church-Turing thesis, which states that any function that is intuitively computable – that is, a function such that we have the impression or the conviction of its being solvable in one way or another, using whatever ideas or techniques that spring to mind at the moment – is also computable with a Turin machine.

In that very same period, Alonzo Church had proposed a different model of calculus (the so-called lambda calculus), which was more formal and less intuitive. Turing, in an appendix to his article, proved that the two models were equivalent.

Gödel, besides being timid and introverted, was also more cautious than the other members of this bunch. When they began to talk to him about these things and of the possibility of constructing a theory that would grasp the intuitive notion of *computable* in a completely general way, he was sceptical. In those years, logic had taken many steps forward, some quite disconcerting. Of some of those steps – crucial ones – he himself had been the greatest and sole artificer. But the results were always tied to a particular formalism, to a specific formal system. This had been the case with the notion of *definable* and with that of *provable*. Why was it necessarily different for that of *computable*? But reflective and honest as he always was, Gödel thought and rethought and in the end, he became convinced that the opposite was true. Once convinced, he was the one who most forcefully underlined the importance of these results every time he returned to the subject.

In 1946 Gödel wrote:

> This importance [of Turing's computability] is largely due to the fact that with this concept one has for the first time succeeded in giving an absolute definition of an interesting epistemological notion, i.e. one not depending on the formalism chosen. . . . For the concept of computability, however, although it is merely a special kind of demonstrability or decidability, the situation is different. By a kind of miracle it is not necessary to distinguish orders, and the diagonal procedure does not lead outside the defined notion.

Here is the miracle that we can present to the petitioner in the case for beatification. It is worthwhile noting that Gödel, in presenting his results, had never before spoken of a miracle.

Again, in 1965 he wrote:

> The concept of "computable" is in a certain sense "absolute", while practically all other familiar meta-mathematical concepts (e.g. provable, definable, etc.) depend quite essentially on the system with respect to which they are defined.

The miracle, then, consists in having formulated a theory that catches hold in an integral way of an intuitive notion – that of *computable* – as well as in the fact that in this theory we are also able to demonstrate several interesting things. Let's mention only two. One positive result and one negative.

The positive result tells us that there exists a *UNIVERSAL* Turing machine, that is, a unique model capable of doing the work of any other particular or specific Turing machine. This is what we are accustomed to today: any computer whatsoever, even our own laptop that weighs less than 2 kg, can do everything (everything, that is, that can be done on a computer; let's not exaggerate!). It's not that my computer can do, in principle, different things than what my friend's computer can do, not taking into consideration concrete limitations such as memory and the like. That is, our computers are – in certain sense – *universal* Turing machines. Now we can better understand what Martin Davis meant in the passage quoted earlier.

The negative result tells us that there are problems that are *undecidable*, that is – roughly speaking – there exist well posed questions to which it is not possible to give "algorithmic" answers. One example is given by the theory known as the "halting problem": there is no algorithm that can tell us whether a generic program which we have given input certain initial values for the variables will – sooner or later – finish running, providing us with the result of the finished computation, or if it will continue to run forever (as might happen if we ask the computer to give us a value for a function at a point where it is not defined). An example that is mathematically filled with this kind of problem is Hilbert's tenth problem.

The theory of computability came out of Turing's head (and other great heads like his). One strange coincidence is that everyone came together in order to find these results independently in the mid-1930s (the works appeared in 1936). With respect to his companions, Turing did something more. We have already mentioned that his model – his machine – could be visualised, in contrast to other proposals, though these were mathematically equivalent. We have also mentioned the determination with which he sustained his ideas. During the war, he was successful in deciphering the secret codes of the German navy and, after the war, he concerned himself with constructing computers while contemporaneously delineating the fundamental mathematical elements of a theory of morphogenesis.

In 1950 he then wrote an article provocatively entitled "Computing Machinery and Intelligence" for the English philosophy journal *Mind*. He jokingly described what computers would be able to do, introducing a game (the imitation game) as an empirical test to establish a machine's intelligence. The day when we can no longer distinguish – from the answers given by a human being and a machine – which is the human and which the machine, will be the day that machines have achieved an "acceptable" level of intelligence (Fig. 1).

In a word, in his free time, he also invented *artificial intelligence*, 5 years before its name was invented. His student, Robin Gandy, who passed away some years ago, recalled that Turing had a lot of fun writing this article, and roared with laughter as he read bits of it to him. More signs of his greatness – the ability to laugh even about his own work, and to have fun while doing important things – absolute greatness.

Gandy, in reconstructing the birth of the theory of computability, noted that the existence of a profound theory is helpful for the development of the technologies related to it. This was the case with electricity, which was based on Maxwell's theory. This has also been the case for computer science, which was based on the theory of computability. But this was not the case for internal-combustion engines, which contributed to the development of thermodynamics instead of finding it already ready and waiting. It is no coincidence that they developed much more slowly.

Up to now we have been lucky with computer science, but the new developments, Internet and distributed systems, have not had a true theory to base themselves on. For the essential problems in these sectors, the theory of computability as a point of reference is too remote or generic to play a significant role. If we want the continued development of our technologies to be carried out quickly, as it has been so far (and not totally alien, as technological development not based on general and profound theory threatens to be), we would do well to invest in fundamental research, inviting everyone to reflect on the important fundamental problems, hoping that sooner or later one of Turing's heirs will lend a hand in providing us with a theoretical point of reference for what is happening.

"Who Is" A. M. Turing

Alan Mathison Turing was born near London on 23 June 1912. Son of an officer in the Indian Civil Service who spent long periods abroad with his wife, Alan was fostered by family friends and attended English Public School, showing talent and a specific interest in following his own ideas, independent of the teaching he received. In spite of this (or perhaps thanks to it) Turing won every single scholastic competition in mathematics.

His first interests and extra-curricular readings concerned Einstein's articles on relativity, then recently published, and the newborn field of quantum mechanics. In 1931 he won a scholarship and entered King's College in Cambridge, where he

Fig. 1 The "Enigma" machine

turned his attention to logic and the philosophy of mathematics, under the influence of Bertrand Russell. He was a sympathiser with the pacifist movement, but he never joined an organisation.

In 1934 he completed his studies and the next year attended an advanced course in the foundations of mathematics taught by Max Newman, with whom he was to remain in contact. During the course he came to know Gödels's theory of incompleteness and Hilbert's problems of decidability, and began to work on his own original approach to them.

He became a fellow of King's College in 1935 with a thesis on the calculus of probability, but he also continued to work on decidability. In 1936 he published the fundamental article "On Computable Numbers with an Application to the Entscheidungsproblem", where he introduced an ideal machine (today called the Turing machine) which formalised the intuitive idea of algorithm beginning with the elementary operations that are characteristic of all calculations.

His work in decidability brought him into contact with Alonzo Church, who at that time was working on the same problems. From 1936 to 1938 Turing studied under Church at the Institute for Advanced Study in Princeton.

Back in England, he was invited to Bletchley Park by the Government Code and Cipher School (GCCS) to participate in the project of deciphering the German Enigma code. Here he was able to put to use all of his skills in logic and statistics joined with his talent for constructing computer machinery. The result was a remarkable contribution to the construction of some "bombes", electromechanical devices for calculating, named for their characteristic ticking. As early as 1941 they were able to decipher the secret messages sent by the German navy. In 1945 Turing received the OBE for his wartime service.

After the war, he took part in a project to construct a computer for the National Physical Laboratory, and returned to academic life in Cambridge and to mathematics. In 1948 he was invited by his former teacher Newman to transfer to the University of Manchester. In 1950 he published another memorable article, "Computing Machinery and Intelligence" in the journal Mind, introducing the topic of artificial intelligence. He was also something of an athlete. He participated in marathons and decathlons, achieving world-class standards.

He was elected a fellow of London's Royal Society in 1951, mainly for his work on decidability of 1936, but his curiosity also drove him to investigate the mathematical structures in biology. In 1952 he published a study on the evolution of living organisms. In the meantime, this being the Cold War period, he had secretly begun to work again for the GCCS.

In 1952 he was convicted of gross indecency for homosexual acts, and as an alternative to prison was sentenced to undergo oestrogen treatment. Because of the conviction, he lost his security clearance and could no longer work on deciphering codes. He and his colleagues and scientific correspondents, both British and foreign, were kept under constant surveillance.

Turing died on 7 June 1954, apparently from eating an apple containing cyanide. The conclusion of the official inquest was that he had committed suicide, but his mother always claimed his death an accident due to carelessness while conducting chemical experiments.

Renato Caccioppoli

Naples: Fascism and the Post-War Period

Angelo Guerraggio

Renato Caccioppoli is probably the most "storied" Italian mathematician, the one who has been most been talked and written about, even beyond the circle of specialists. He has been made familiar as a personage to a vast public (though his research topics have just been touched on) in an attempt to accomplish the difficult task of communicating how complex and fascinating mathematical thinking is. (Although is still hard to do, we can no longer complain about the unfavourable conditions or lack of opportunity for popularising mathematical methods and ideas.)

Much has been said about a whole series of meetings and conferences organised by his mathematical colleagues in Naples, but in other cities as well, which just goes to show that the memories and affection that tie Caccioppoli to his native city find echoes of interest and generosity in other research communities as well. There was the congress in 1968, that of Pisa in 1987, and then the "days" in Naples in February 2004, and then that of the following April in Rome (organised by the *Institute for the Applications of Calculation* (IAC) and the *National Research Council* (CNR)). And of course we can't not cite the 1992 film "Morte di un matematico napoletano" directed by Mario Martone, with all of the discussion and debates that it gave rise to: television programs, books, biographies, commemorations, interviews that spoke about Caccioppoli and Naples in the 1950s. And then there was the novel *Mistero napoletano* by Ermanno Rea. There are no end of anecdotes, always charming, always told with affection.

We have reason to celebrate the life of Caccioppoli – even more than his tragic death –, and the life of a great mathematician is found above all in his research. Trying to summarise 30 years of work in just a short space inevitably leads to some arbitrary choices, but there are some points that are sufficiently "stable" to give a first, brief idea of Caccioppoli's contributions:

- The first papers, around 1926, on the *extension* of the definition set of a linear functional using the technique of *extrapolation* that would characterise later works as well, and would find an immediate application in *integral theory*.
- Studies on a *geometric theory of measure* for a surface defined parametrically, which took into account the work of Lebesgue (as well as the more recent papers

by Banach and Vitali) and which led him in the years 1927–1930 to consider oriented surfaces and of the dual attributes – of *extension* and of *orientation* – to the area element; these studies would be taken up again in 1952, with the baton being passed to Ennio De Giorgi.

- The studies beginning in the 1930s on *ordinary differential equations* (including the generalisation of an existence theorem of Bernstein concerning among others a limit problem of a second degree equation) and *partial differential equations*, particularly elliptic: an existence theorem within the class of functions whose second derivatives are Hölder; various a priori upper bounds; the proof that C^2-class solutions of analytical elliptic equations are analytical, with the first answer to the nineteenth problem posed by Hilbert at the 1900 International Congress of Mathematicians, etc.
- The "discovery" of functional analysis and the fixed-point theorems at the beginning of the 1930s; the limited applicability of the theorems concerning the solutions of the equation $x=S[x]$ to differential and integral equations would then lead to the formulation of the *principle of functional correspondence inversion*, the result of considering the transformation $T[x]= x - S[x]$
- Studies on the *functions of more than one complex variable*, and on analytical and pseudo-analytical functions.

Renato Caccioppoli

Thanks to these studies and others, Caccioppoli undoubtedly deserves credit for having carried Italian analysis to the most advanced fronts of research. Carlo Miranda wrote, "it is above all thanks to the courses he charted that Italian analysts were able to overcome the isolation they experienced during the war years and those immediately after without too much harm having been done". Caccioppoli's "modernity" – with respect to what was going on in the international arena at that time – could also be evaluated indirectly, by means of the disputes and controversies over priority posed by his papers. The names Dubrovskij (for functions with limited uniform variation or uniformly additive), Radó (for controversies over measure theory), Stepanoff (for asymptotically differentiable functions), Petrovsky, Perron and Weyl (for the lemma on the harmonicity of functions orthogonal to any Laplacian) testify to Caccioppoli's activity in the research of the day, destined to leave profound marks on the history of analysis in the twentieth century.

The process of reorientation towards the most promising research contents particularly holds for functional analysis.

As a discipline, functional analysis was born during the final decades of the nineteenth century, a development of the driving ambition, which would reach maturity over the course of the following century, to address not only numeric or geometric problems, but problems of any nature whatsoever – whatever their content, and whatever kind of object involved – by means of the set of such objects and their structure. This project was of immediate interest to mathematicians of the like of Salvatore Pincherle, and above all, Vito Volterra. It was Volterra who formulated the precise definition of the concept of a functional, or *line function*. He also developed a calculus of functionals, analogous to ordinary calculus, starting with the definition of a directional derivative which would later take the name *Gâteaux-Lévy derivative*. Fréchet's 1906 *thesis*, entitled "Sur quelques points du calcul fonctionnel", marked a turning point in the development of the discipline. Volterra's contributions to it were motivated by the need for new, more refined instruments that could be *applied* to problems of mathematical physics or to other questions which had progressively become part of the mathematical research itself (such as in the case of integral equations).

Starting with Fréchet, functional analysis increasingly became a discipline in its own right, scarcely needing to justify its developments for possible applications. It usually happens that it is the successive generation – the first students of the "founders" – who, finding themselves with new *instruments*, formulated as such, study their natures and transform them into the central objects of an autonomous theory. This didn't happen in Volterra's case. His choices – in mathematics, as well as in life – prevented him from creating a genuine school, and he continued to think of functional analysis as a field of research that was to be developed in a climate of "relative" freedom, always with an eye towards the so-called *applications*. As evidence of this we need only recall his gentlemanly but obstinate quarrel with Fréchet regarding the definition of the derivative of a functional. In Volterra's case it is perhaps more correct to speak of *non-linear analysis*; or perhaps, of an Italian way of doing functional analysis, at a time when Fréchet and Moore were decidedly oriented towards a more *general* way of doing it. The fact is that at the 1928

International Congress of Mathematicians in Bologna, Fréchet promptly gave a snapshot of the then current situation: *si, en Italie, l'Analyse générale proprement dite n'a pas encore trouvé d'adeptes, n'oublions pas que cette science nouvelle est née de l'Analyse fonctionnelle, merveilleuse création du génie italien* (if in Italy, general analysis proper has not yet found any supporters, let's not forget that this new science was born of functional analysis, a marvellous creation of Italian genius).

It was to a large extent Caccioppoli who filled the gap, in the 1930s. In the period between the two world wars, hardly anyone else published articles about functional analysis *à la Banach*, if you will; perhaps the only other one is only a 1932 article by Guido Ascoli on metric linear spaces. As we said, Caccioppoli worked on *fixed point theorems*. He published a series of three articles on this subject: "Un teorema generale sull'esistenza di elementi uniti in una trasformazione funzionale" (1930); "Sugli elementi uniti delle trasformazioni funzionali: un'osservazione sul problemi di valori ai limiti" (1931); "Sugli elementi uniti delle trasformazioni funzionali: un teorema di esistenza e di unicità e alcune sue applicazioni" (1932). In the first he proved an existence theorem "of a topological nature" for the fixed points of a continuous transformation on $C[a,b]$ even when their domain could be given equivalently – Caccioppoli added – by either $C^n[a,b]$ or $L^2[a,b]$. He then states the theorem of existence and uniqueness for a contractive function in a complete metric space (including the algorithm of convergence, simply remarking that "the proof is obvious"). This was the theorem proven by Banach in normed vector spaces in his 1920 thesis and then published in *Fundamenta mathematica* in 1922, just as the first applications of the fixed point theorem "to the study of functional equations" were due to Birkhoff and Kellogg. In the second of his three articles (the third one marks the passage to the inversion principle of functional transformations), Caccioppoli candidly admits that he "takes advantage of the occasion to acknowledge the priority of the authors". This timely admission – Schauder had published his work on fixed point theorems just shortly before – would confirm Caccioppoli's importance for the history of analysis in Italy: some credit for priority is due to him, and some should be taken away, but on the whole, these show that Caccioppoli was sailing in high seas and did not restrict himself to the safer "territorial waters" of a national tradition.

Caccioppoli's contributions were remarkable. It's not only a question of the "number" of theorems proven, but is the constant attempt to "think big". Carlo Miranda wrote that Caccioppoli "didn't love honing and polishing". Using other words but with an analogous meaning, the commission of the competition of Cagliari (which designated him as the winner) wrote that "the orientation of his research is predominantly critical". Caccioppoli did not limit himself to manipulating notions that had already been defined, but tried to develop general theories. The metaphor of Grothendieck comes immediately to mind, when he said that thought can choose to live in a house that is already constructed by preceding generations, perhaps moving some walls and adding a porch, and that this search for comfort – this more or less painstaking do-it-yourself work – constitutes the heart of a certain kind of academic mathematics. But thought can also choose to live in unexplored territory, where it slowly constructs its own house. In Caccioppoli we see the latter choice, where he is always trying to find the right concept, with an optimum degree of generality. In his articles, which perhaps report on a given study but still express a general orientation – we find passages that speak of the "naturalness" of a formulation or of a "natural field of existence of the functional": "the generality of the hypotheses often assure not only the generality but also the simplicity and the coherence of the results obtained"; or, "the problem of squaring surfaces has to be solved with the same degree of generality as that of the rectification of curves". He is instead critical of other generalisations; though correct, they turn out to be "laborious and seem to march towards ever greater complication", or "they break up the primitive unity of the theory", or they are "fragmentary and reciprocally unrelated".

Obviously, this "thinking big" also means taking bigger risks, even for Caccioppoli. If you read the two volumes of his collected works, *Opere*, published by the *Unione Matematica Italiana* in 1963, you will find that there often appear – on the part of the editors – phrases such as, "the function should be substituted by...", "the proof of this theorem is not exact", "the hypothesis is not always explicitly stated". In one case, Caccioppoli himself spoke of an "oversight" of his that Tonelli had noticed, and added somewhat ironically, in regard to his first papers on the geometric measure theory, "if some of the ideas (but not all) that inspired my work are today rather widely known, some of the errors contained in them have either been ignored or quoted ... This carelessness is certainly deplorable but might be said to be *felix culpa*, if it hasn't prevented the discovery of essential facts or more suitable methods".

To describe Caccioppoli's mathematical personality, it should be noted that he was part of a generation that believed deeply in research. Mathematics is not just a profession. It is a cumbersome and tyrannical taskmaster. Working 24 h a day and mixing up days and nights – as Laurent Schwartz described Grothendieck's life – is not simply the result of a "career" choice. It is the expression of the awareness that one is in possession of privileged instruments that make it possible to understand, know and transform. It is possible to dedicate one's very life – not just professional life, but one's whole existence – to such concepts and instruments.

Those who knew Caccioppoli easily recognised his status as an outstanding researcher. In the world of mathematics, where the term *genius* is not lightly tossed around, it was soon attached to Caccioppoli. But the life of the *mathematician* Caccioppoli, who strongly believed in the value of mathematical culture, was never one of only scientific research.

Caccioppoli always lived in Naples, with the only exception being the period from 1931 to 1934 when he taught in Padua, substituting Giuseppe Vitali, who had transferred to Bologna. In the cultural circles of Naples, he was know for his passion for music and his bravura as a pianist (and also as a violinist). Also legendary were his love and comprehension of contemporary French literature, with a special passion for Rimbaud and Gide. After the war, his love for cinema led him to organise a group called *Circolo del cinema* in Naples; the Sunday morning films and Caccioppoli's presentation of them were a standing appointment for many fans of cinema.

But before we get to the post-war years, we should take a look at the years of Fascism. Caccioppoli was a firm opponent of the regime. His ironic dissent using the "rooster on a leash" is well known: when the Fascist party advised men not to walk dogs because it was considered not very masculine, Caccioppoli walked down Via Caracciolo with a rooster on a leash. Much more serious was the episode involving the "Marseillaise", which Ermanno Rea described so well in *Mistero napoletano*. At the beginning of May 1938, Hitler was about to arrive for a visit to Naples. Caccioppoli and his future wife, Sara, went into a beer hall late one evening and, annoyed by a group of Fascists singing "Giovinezza", the official anthem of the Italian National Fascist Party, he sat down at the piano and sang the French national anthem, the "Marseillaise", at the top of his lungs. He was immediately arrested. Punishment was really severe for pranks like this. In order to save him from being thrown into prison, his family claimed that he was mentally ill, and he was admitted to an insane asylum rather than prison.

Although this is how the story is told, the official police reports actually paint a grimmer picture. First of all, they record the episode as having taken place on 23 October 1938, so Hitler doesn't have anything to do with it. Then, rather than a beer hall, they say that it took place in a local pub "frequented by persons of modest extraction", "a tavern located in Naples on the Riviera di Chiaia", where a man – Renato Caccioppoli – "of decent aspect" is however described in another report as "shabbily dressed" or "badly dressed" and a woman – elegant, spirited, lively and who spoke French to her companion (who pretended to be Russian) "of an easy nature and with quite liberal manners" – "after having drunk some wine, offered another round to a group of labourers who were in the tavern. The two individuals fraternised with the labourers, and then left with them after they had finished dancing". There is no mention of the Marseillaise in the police report. The substance, however, remains: "offered wine in return for pizzas . . ., political conversations with the labourers . . ., slurs about Italian politics (in comparison to the French) that continued on the funicular bound for the Vomero". Then the arrest, described by the federal secretary of the Fascist party: "by virtue of the authority of Public Security their arrest was immediately effected. Caccioppoli, during the final interrogation, showed signs of mental imbalance and thus, after having been examined by a psychiatrist and diagnosed as insane, he was admitted to an insane asylum". The police report uses the same tone, "said person having shown signs of mental imbalance during the course of interrogation. . . he was diagnosed as demented".

Caccioppoli's anti-Fascism was well known. Earlier the police in Padua had put him under "suitable political observation", even though in a document dated August 1933 it was admitted that "given the subject that he teaches he certainly cannot use it to make his ideas known, but with his close friends he expresses himself violently against anything that has to do with Fascism". In Naples, there is no doubt among those in the police force: "Caccioppoli, aside from his unassailable value as a scientist, because of his immoderate use of alcohol in his private life, shows himself to be abnormal and without any social values". After the episode of the "Marseillaise" – if indeed there ever was such an episode – the newspaper of Italian expatriates in Paris, *La voce degli italiani*, ran an article under the headline "Prof. Caccioppoli arrested, tortured, driven mad" in which it was reported that the mathematician was "tortured so severely that he is currently in a mental hospital". On 25 April 1939, the rector of the University of Naples, requesting an extension of Caccioppoli's leave of absence from the Ministry for Education, wrote,

> Prof. Caccioppoli is considered to be affected by an imbalance (which we hope is quite temporary) and thus does not possess full control of his faculties and cannot adequately perceive and evaluate the various contingencies and occasions of social life, a condition often found in those whose intelligence has taken over and who, completely absorbed in the study of arduous disciplines that require intellectual polarisation and particular dedication, are almost completely estranged from the rest of life's circumstances. With regard to this, the recent case of Prof. Maiorana naturally comes to mind. And it is neither out of turn nor beside the point to recall that these are young men, indeed almost lads, who were given university chairs at an age in which other young men are only in the middle of their education, and thus they found themselves in positions of responsibility as university professors, completely unprepared to face the studies and the requirements of an environment from which they had been completely estranged during the period of relentless and all-absorbing studies which led to their being given the chair.

The episode of 1938–1939 was not the only time that Caccioppoli was arrested. A similar thing happened in 1952, this time the work of the police of the Italian Republic. It was during the post-war period. Caccioppoli had worked in favour of the Republic during the 1946 Referendum. Later, he formed closer relations with the communists in Naples, the only viable alternative to the crudeness and superficiality of the supporters of Achille Lauro. He was a faithful supporter of the Italy's Internationalist Communist Party, although he never officially joined. He was involved in the events of the *Gramsci group*. He joined the "peace partisans". It was as a pacifist and firm opponent of American intervention in Korea that he was arrested on 16 June 1952. The official report this time says:

> On 16 June 1952, the day before the arrival in Naples of Gen. Ridgway, Prof. Caccioppoli ... having gathered about 200 students from his course and others led them to the central building of the university, where he gave a speech protesting against the aforementioned general's visit to Italy and in favour of peace. This speech gave rise to a vehemently hostile protest in the form of invectives thrown in the direction the American seamen who were housed in the hotel in front of the university and against American automobiles passing by.

The reaction of the Minister for Public Instruction, Antonio Segni, lead to Caccioppoli's being severely reprimanded for having incited the disturbances that followed his speech, and for behaviour that constituted an obvious "infringement of the disciplinary rules of the University".

Traces of Caccioppoli's political and pacifist leanings can be found in his correspondence with Mauro Picone, housed in the archive of the IAC, the *Institute for Applications of Calculation* founded by Picone in Rome, and recently published in *PRISTEM/Storia* (no. 8/9, 2004).

In a letter dated 11 August 1953, Caccioppoli wrote that,

> ... idiotic problems with the police have forced me to renounce going to Poland. Can you believe that after weeks of stalling, they gave me back a ... passport that had been annulled for all countries (even France!) but ... extended for Poland and "countries of transit" (?)

until 6 September, *the opening day of the congress*. This after having transcribed all of the information from the telegram inviting me. With this kind of "passport" it would be hard to get past Tarvisio. To add insult to injury, it permits "one trip only"!!

In a letter dated 20 August 1958 he mentions a demonstration that took place in Naples:

> In effect, I didn't take part in the peaceful demonstration in Via Roma that provoked the usual hail of blows by riot police with batons, unannounced and equitably distributed between demonstrators and simple passers-by. I followed the trial because it interested me politically and because among the principle defendants were some of my best friends.

The correspondence is interesting for more than this, however, primarily because of the fact that it sheds light on the relationships between the two mathematicians. Picone (1885–1977) is known in the history of Italian mathematics for some valuable works dealing with partial differential equations, but above all because he founded a school that produced some of the greatest Italian analysts of the second half of the twentieth century. The instrument that led to the founding of this school was the INAC, the *National Institute for Applications of Calculation*, later renamed simply IAC, or *Institute for Applications of Calculation*.

Founded in Naples in 1927, and then transferred to Rome a few years later as a part of the National Research Council, the IAC soon became a significant new presence on the horizon of Italian (and not only) mathematics. It represented a new numerical mentality. It was no longer sufficient to prove an existence theorem, or even one of uniqueness, but it was necessary to outline the procedure for effectively calculating the solutions. In other words, it required that the same attention and the same rigour be applied for determining the numerical algorithm, the proof of its convergence and the upper bound of the error of approximation. The objective was the synergy with applications for experimental disciplines, for the study of "their" mathematical problems, and the numerical determination of the solutions. It was the first time that mathematical research had been organised outside of the closed academic circuit. It was the first time that young people could be started on a path that led to a considerable number of possible jobs. It was the first time that mathematics became a subject and object of consulting, opening new professional relations and giving rise to team research. It was the beginning of a road that would culminate in the UNESCO conference in Paris in 1951, which nominated Rome and the IAC as the headquarters of the *European Centre for Calculation*.

That Picone was sympathetic towards Fascism is a known fact, and was evident long before the need to support and manage the INAC led him to curry favour with political authorities. He himself said that he was "a black shirt from the very beginning". On 5 June 1923 he wrote to Giovanni Gentile, who had just joined the Fascist party,

> Your illustrious and venerable Excellence, permit me to express my most heartfelt pleasure at Your Excellence's joining the National Fascist Party, of which I too am a member. This newest member of the fascist party – so prominent – and the considered statements contained in that letter, will overcome the hesitancy of many Colleagues and bring new, pure blood into the robust veins of the party that will reconstruct and renovate the Nation!

Not even later would Picone deny his enthusiasm for the Fascists, not would he express any self-criticism or try to distance himself from the two decades of Fascist policies, except for one brief exception, when, years later commemorating Terracini, he spoke of his "painful exile in Argentina".

Given this situation, the correspondence between Caccioppoli and Picone brings quite a few surprises. Caccioppoli – "communist" and deeply committed to a democratic system – had no problem in continuing to correspond with the "fascist" Picone, in a post-war period that in any case was characterised by opposing sides in strong contrast. Indeed, his letters to Picone are quite sincere and infused with a deep affection and heartfelt esteem. He never refers to previous – and embarrassing – political positions supported by Picone. On the contrary, in some way he wants to help to put them in a proper perspective, reducing them to a pragmatism that is inevitable "pro-government": "you don't get involved in politics, I know, and maybe, devoted as you are only to your work, you may be willing to tie the donkey up where the owner wants you to; but not me" (letter dated 19 July 1954).

The first element that emerges from a reading of the letters is the almost filial affection that Caccioppoli shows towards Picone – expressed, naturally, given his temperament, without any sugariness or fawning. And Picone surely returned both the affection and esteem. He writes of "a great mathematician who, alone in Italy, is master, by dint of critical sense as well as invention, of the foundations and the advances, of all three of areas of analysis, topological, real and complex, as well as of their applications and concrete problems". He doesn't hesitate to declare more than once that the student had surpassed the teacher. And what lengths he went to in the effort to make sure that the merits of that student were acknowledged by the scientific community! There was the time in 1951 that he tried to ensure that Caccioppoli was awarded the *International Feltrinelli Prize*, followed by a second attempt in 1956, as well as the campaign to have him elected a *national member* of the famed *Accademia dei Lincei*: "instead of receiving honours, he has for some time now been persistently subjected to the most vulgar and unjustifiable slur campaign on the part of some quite reputable mathematicians. Here ... the *Accademia dei Lincei* has the obligation to intercede, since they are above all the fervid upholders of the nation's values, disavowing those vulgar denigrators". And how furious he became with Gianfranco Cimmino, another one of his favourite students, when in a first draft of the preface to the collected works of Caccioppoli his role in the formation of the mathematician was neglected:

I must however complain about the complete absence in your preface of any mention whatsoever of the influence that I undoubtedly had in orienting Renato early on towards studies of functional analysis and the modern foundations of the theory of functions of real variables that it is based on. This influence is undeniable and can be proven. When in long ago 1925 I arrived at the University of Naples I found Renato in his third year of university in the throes of his thesis on Pfaffian systems, disgusted with mathematics and undecided as to whether to continue studying it or to change to a career as an orchestra director. He attended my classes in higher analysis in which I discussed Lebesgue's integration theory and I remember quite well that he showed me what he had in him during a lesson in which I assigned my students the task of finding an example by virtue of which a hypothesis formulated about a certain theorem was shown to be essential. When I finished the lesson

I was chased by a shaggy young man, shabbily dressed, who stuttered out that he had found the example I had asked for. I invited him to come into my office and he showed me a very elegant example that completely fulfilled the conditions I had set. This was Renato, and we had a long conversation. I sensed his powerful genius right away and from then on I was tied to him by a friendship that was never to dim. We began to see each other almost every day and I talked to him about modern functional analysis and its applications to problems of the integration of differential equations. His mother learned about this friendship and came to see me one day to tell me how grateful she was for the interest I had shown towards her son, who, to her great relief, appeared to have given up the idea of abandoning mathematics to become an orchestra director. She also told me that Renato called me the "Stravinski of mathematics". Naturally, when Renato set out on the new path that I had opened for him, he made giant strides, and I have to admit that within a short time our roles were reversed, that is, he became the master and I the disciple. But by God, I swear this is the truth: I was the one who saved dear and much mourned Renato's formidable genius for mathematics.

Another element that explains why the relationship between Caccioppoli and Picone continued uninterrupted and cordial in spite of their many differences is that they both belonged to the mathematics community. Perhaps in this case, for this generation and for Caccioppoli in particular, the term *community* is not unfitting. It is not rhetorical, nor does it refer to a merely sociological fact. It means a common rationality, a common sensitivity, and common values – to be sure, not shared by all mathematicians. Caccioppoli was quite far from an indiscriminate appreciation of his colleagues; in fact, these values came to count almost more than did political leanings (to which even Caccioppoli was quite tied). On the other hand, this was the very stance that he had taken during the time of the dismissals in the convulsive stages that followed the events of 25 September. The steps taken by the rector Adolfo Omodeo, president of the *Commissione per l'epurazione*, the Commission for Purification, affected – among the mathematicians – only Giulio Andreoli. The letter notifying him of his dismissal, dated 7 October 1943 (all of the professors in question were reinstated during the summer of 1945) contains an implicit reference to the case of Gaetano Scorza: "You were always an accomplice of the Fascists. . . . For years you have taken aim at students and colleagues for political reasons, and sometimes so harassed some famous professors of the department of mathematics of the University of Naples that they were forced to request transfer to other universities". Caccioppoli was naturally on the side of Omodeo and the Commission, but his letter dated 15 March 1944 betrays a certain lack of enthusiasm for acts that will inevitably strike some colleagues: "It would seem a refusal on my part to recognise your efforts aimed at restoring the liberty and dignity of our university if I were to deny my concurrence in a part of it that is as necessary as it is painful".

What Caccioppoli could not bear – be it in the case of the supporters of Fascism, or the case of the supporters of Achille Lauro, although he is more indulgent towards his mathematical colleagues – is the arrogance of ignorance, that is, the union of the two. I don't think that we can speak of snobbery in Caccioppoli's case: he was an intellectual who, through nearness to and keeping company with the Italian communist party, is still legendary with the working class and who – in actual fact, during the war years – gave his all to organising a strike of transportation workers. His is rather the decided aversion to what Gerardo Marotta identified as the lower middle class's most aggressive and vulgar instincts and its pettiness. He simply could not live with a milieu that is so apathetic and intellectually lazy.

This is the the theme that provides the key to his solitude. Genius or immoderation? The mad mathematician? Beyond some consequences for his temperament and of a tendency – this too, in any case, not at all natural – towards isolation and solitude are the aspects that progressively emerge in his life. And too, these explain his "attachment" to Picone. Just think for a moment of the outcome of his marriage to Sara who, we recall, shared in the experience of the "Marseillaise" (and also shared pizza and wine with a group of workers). These are feelings that emerge in the letter to Picone dated 19 July 1954 that was quoted from earlier:

> I wrote you some months ago that I would *not* be going to Amsterdam, explaining why. You answered me, saying that you "don't accept", which I took as a impulsive show of your generous temperament which, believe me, no one appreciates more than I do. But if anything your "don't accept" should have been said to the Scelbas, or the Fanfanis, or to any of the many Italians we can recognise them in, and not to me, who, like so many, if not actually an "enemy of the nation", are at least among those citizens who are discriminated against, that is, those who don't enjoy the full rights guaranteed by the constitution. The borders of our "liberal" nation can be crossed by a [word missing] acknowledged drug smuggler but not by Prof. Renato Caccioppoli, suspected rightly or wrongly of smuggling ideas.

These feelings of solitude and others were his unhappy companions up to the tragic event of 8 May 1959.

Bruno de Finetti

The Foundations of Probability

D. Michele Cifarelli

It is impossible to try to describe Bruno de Finetti's work within the limits of a few pages, so numerous and significant are the contributions he made to diverse areas of mathematics (applied and non), as well as other branches of knowledge such as economics and biology, to mention only two.

Bruno de Finetti was born in Innsbruck on 13 June 1906 to Italian parents. In 1923, at the age of 17, he enrolled in the Politecnico di Milano but by the time he reached his third year, he changed directions and enrolled in mathematics at the University of Milan, where he earned his degree in 1927 with a thesis on geometry with Giulio Vivanti as his thesis adviser.

But even before his transformation into a mathematician, inspired by a work by the biologist Carlo Foà, de Finetti delved into some research in population genetics, which was the subject of the first of his numerous works published in 1926 [1]. This was the first example in the literature of a model which took into account several superimposed generations, thus anticipating research in population genetics by some decades.

Immediately after receiving his degree in applied mathematics, de Finetti accepted a position at the Central Institute for Statistics, at the time directed by Corrado Gini, the institute's founder.

De Finetti remained with the institute until 1931. These were the years in which the foundations were laid for his primary contributions to probability theory and statistics. With the subjective approach, probability appears as a measure of the observer's belief that a given random event will occur. De Finetti was not himself aware that the same approach had been conceived by F. P. Ramsey in 1926; but in any case his own emphasis was placed on a coherent assignment of probability, instead of on rational decisions as Ramsey had.

As a consequence of subjectivism, statistic inference was no longer seen as an empirical process that arose only from the available data, but rather as a logical process capable of producing "opinions" about the predictions that were compatible with the available data (see his works [2–6]).

It was during these same years that de Finetti introduced the notion of "exchangeable sequences of events" and refined his analysis to arrive at the celebrated representation theorem, which says that any law of probability relative to (infinite) sequences of exchangeable events can be represented by a mixture of laws of probability of independent events with the same probability of success.

Bruno de Finetti

It was the introduction of this fundamental notion (which was, to be truthful, only defined by the mathematician J. Haag at the International Congress of Mathematicians in Toronto in 1924 and published in 1928) that made it possible for de Finetti to justify the determination of a (subjective) probability by means of a frequency and to reconstruct the Bayes–Laplace paradigm, eliminating from it equivocal statements such as that, common up to that point, of "independent events of constant but unknown probability".

In 1929, de Finetti also began a study of independent increment processes. The crisis of determinism and the principle of causality had constituted fundamental innovations in scientific method, with probabilistic reasoning replacing classical logic. De Finetti's pioneering research in random functions served the precise purpose of translating deterministic laws into laws with elements of probability (see [7–9]). It was no coincidence that he deduced, from the general results that had been established, the laws of probability of some functionals of the well known process of Wiener–Levy, as well as (in [10]), the description of the laws of probability that are infinitely divisible, which then became the point of departure for Kolmogorov's and Levy's research regarding these laws, culminating in the so-called representation theorems of infinitely divisible laws.

It's nice to think that de Finetti came back to the theme of independent increment processes in a paper of 1938 [11], in which he assumed a position that was critical of his own work, and in particular regarding the considerations of propositions that are "transcendent", or infinite. For de Finetti, only "objective" propositions lend themselves to probabilistic evaluation.

As is well known, de Finetti believed that probability is equivalent to a wager that a random event will occur. In this sense, probability is in many ways analogous to the functional approach to the *mean*: both are based on the primary notion of there being no difference between certain and uncertain results and a set of rational requirements. The notion of *means* was addressed in a work by de Finetti in 1931 [12], where he took as a point of departure the definition of mean that had been given by Oscar Chisini 2 years previously, generalising it for the case of distributions (of probabilities, frequencies, or other) and arrived at a description of an *associative mean* (as far as we know, the first appearance of the adjective in the literature), thus establishing what has come to be known as the *representation theorem of associative means*, and which goes by the name de Finetti–Kolmogorov–Nagumo. Andrey Kolmogorov and M. Nagumo came to be associated with this important result because in 1930 each of them proved, independently, a particular case of this theorem.

In 1931 de Finetti moved to Trieste, where he had accepted a position in actuarial work with the large insurance firm Assicurazioni Generali. During his time in Trieste he developed the research he had begun in Rome and achieved significant results in financial and actuarial mathematics as well as in mathematical economics. He was also very active in the mechanisation of some actuarial services, which led – in all probability – to his becoming one of the first mathematicians in Italy who was able to solve analytical problems by means of a computer.

In three papers presented to the Accademia dei Lincei [8–10], and especially in a very famous work of 1937 [13], de Finetti extended the representation theorem of exchangeable events to the case of exchangeable random variables, which he stated without having a complete hypothesis of additivity at his disposal. In the modern formulation the theorem appears in the literature with the hypothesis of σ-additivity and goes like this: the succession of random variables $(x_n)_{n \geq 1}$ is exchangeable if and only if there exists a measure of random probability p such that, given p, the random variables of the sequence are conditionally independent according to that law of probability p. This was the result that paved the way for the solution to problems that are today on the cutting edge of statistics, called "non-parametrical inferential problems".

De Finetti devoted one of his papers [14] to the "player's ruin", following Lundberg's point of view. This contribution, in addition to introducing some notions typical of the world of insurance (such as the *level of risk* for initial capital), contains connections to an interesting identity discovered some years later by Wald.

De Finetti was an authoritative participant in the discussions that followed the formalisation of the expected utility theory (Daniel Bernoulli's principle) by von Neumann and Morgenstern. However, the attention it deserved was not given to his important contribution of 1952 [15] in which he presented the measurement of risk aversion, later known by the name Arrow–Pratt. This measure, introduced by de Finetti as well as by Kenneth Arrow and John Pratt (1964) arose out of a consideration of lotteries whose possible outcomes are only slightly different from each other, and thus with a local analysis of risk aversion, and defining risk aversion as the deciding factor so that the value of the lottery is less than its average value. This then involves conceiving an index of concavity of the utility function aimed at measuring risk aversion. De Finetti indicated it with the ratio $-u''/2u'$, where u is the utility function, and his justification of it was much more detailed than that later given by Arrow and Pratt. Yet again de Finetti had anticipated future analysis and results!

It was also in Trieste that de Finetti began his academic career, teaching courses in financial mathematics, probability, and mathematical analysis. Only in 1947 was he given the chair in financial mathematics at the University of Trieste, although he had won the competition for that position in 1939. In 1954 he transferred to the Faculty of Economics at the University of Rome, and in 1961 he moved to the Faculty of Sciences, where he remained a professor of probability theory until 1976. He had scientific contacts with a great number of other scholars in Italy and abroad. In particular, he had the opportunity to meet many eminent mathematicians working in probability and statistics, among them F. P. Cantelli, G. Castelnuovo, M. Fréchet, A. Khinchin, P. Levy, J. Neyman, R. A. Fisher, G. Pólya and J. Savage. With Fréchet, even before his main publications on the foundations of probability, he carried on an important correspondence, which had grown out of his paper [6]. For a discussion of this and other elements of de Finetti's work, the reader can read the paper by D. M. Cifarelli and E. Regazzini, "De Finetti's contribution to Probability and Statistics" in *Statistical Science* (1996). For a complete bibliography of de Finetti's work, see "Bruno di Finetti" by L. Daboni (*Bollettino dell'Unione Matematica Italiana*, 1987).

De Finetti's innovative ideas regarding probability were thoroughly and masterfully presented in the two-volume *Theory of Probability* translated into English in 1975 from the original 1970 publication in Italian by Einaudi.

At the time of his death, on 20 July 1985, de Finetti was an honourary member of the Royal Statistical Society, a member of the International Statistical Institute and of the Institute of Mathematical Statistics. In 1974 he was elected corresponding member (later became a full member) of the Accademia dei Lincei.

Works by Bruno de Finetti Cited

1. *Considerazioni matematiche sull'ereditarietà mendeliana*, Metron, 1926.
2. *Probabilismo, Saggio critico sulla probabilità e sul valore della Scienza*, Logos Biblioteca filosofica, Perrella, Napoli, 1931.
3. Sul significato soggettivo della probabilità totale alle classi numerabili, *Rend. Reale Ist. Lombardo di Sc. e Lett.*, 1931.
4. A proposito dell'estensione del teorema della probabilità totale alle classi numerabili, *Rend. Reale Ist. Lombardo di Sc. e Lett.*, 1930.
5. Ancora sull'estensione alle classi numerabili del teorema della probabilità totali, *Rend. Reale Ist. Lombardo di Sc. e Lett.*, 1930.
6. Sui passaggi al limite nel calcolo delle probabilità, *Rend. Reale Ist. Lombardo di Sc. e Lett.*, 1930.
7. Sulle funzioni ad incremento aleatorio, *Atti Acc. Nazion. Lincei*, 1929.
8. Sulla possibilità dei valori eccezionali per una legge ad incrementi aleatori, *Atti Acc. Nazion. Lincei*, 1929.
9. Integrazione delle funzioni ad incremento aleatorio, *Atti Acc. Nazion. Lincei*, 1929.
10. Le funzioni caratteristiche di leggi istantanee, *Atti Acc. Nazion. Lincei*, 1930.
11. Funzioni aleatorie, *Atti del 1° Congresso UMI*, 1938.
12. Sul concetto di media, *G. Ist. It. Att.*, 1931.
13. La prévision: ses lois logistiques, ses sources subjectives, *Ann. Insti. H. Poincaré*, 1937.
14. Teoria del rischio e il problema della rovina dei giocatori, *G. Ist. It. Att.*, 1939.
15. Sulla preferibilità, *Giorn. degli Econ.*, 1952.

A Committed Mathematician

Gian Italo Bischi

De Finetti's commitment to mathematics education was always concrete and vigorous, as shown by his publications of treatises, textbooks, educational papers, and articles for non-specialists, as well by intense organisational activities.

He was the president of the Associazione Mathesis from 1970 to 1981, and during that same period he was the director of the *Periodico di Matematica*, where he published numerous contributions in which he strongly supported the need to make mathematics intuitive, a position opposite to that of the Bourbaki group regarding mathematics education. In 1962 in Rome he began the first mathematical competitions among students, in the wake of similar earlier experiments already undertaken by Giovanni Prodi in Trieste, later developed under the auspices of the Club Matematico, founded by Giandomenico Majone in 1964 to promote seminars on problems of mathematics education.

He came out numerous times against the situation facing mathematics teaching in Italy, sometimes even in ways that were provocative and ironic, as can be seen by this following extract, referring to the written examination in mathematics at the high school for scientific studies:

> What is involved is an example of unsurpassed pathology of an aberration intended to foster the systematic and total stultification of young people ... Since time immemorial (at least for some decades) it has been the case that in this notorious written examination the very same stereotypical problem was repeated precisely, with only a few variations (second-degree equation, or trinomial, with a parameter; hence we could use the term "trinomitis" to indicate the excessive insistence on this particular argument alone): a problem that above all has the misfortune of being reducible to a scheme that is mechanical, formal and pedestrian, and which carries the name of a certain Tartinville. As for myself, I came only lately to learn about and despise Trinomites and Tartinvillites: I didn't take seriously the remarks, negative but expressed to me in general terms, about mathematics in the high schools for scientific studies expressed by some colleagues at the time I chose it for my daughter.

C. Bartocci et al. (eds.), *Mathematical Lives*,
DOI 10.1007/978-3-642-13606-1_16, © Springer-Verlag Berlin Heidelberg 2011

Another example is the following remark at the C.I.I.M. congress held in Viareggio in October 1974:

> Any appropriate and considered choice on the part of the teacher is rendered impossible and inconceivable by the whole supporting structure of rules inflicted, in Italy, on the University (as in all Schools and on Public Administration in general), rules which can aptly be called bureauphrenic (in France they also use an even cruder term "bureausadistic") and juridiculous (a synthesis of the two terms overlapping by three-fourths, juridical and ridiculous).

Finally, mention must be made of de Finetti's commitment to questions of politics and economics. He was ever an attentive and critical observer of social facts, which he analysed with the scientist's purity of reason, often making evident how twisted and unjust things were, and upholding the importance of individual liberties and democracy. In the context of university life, he believed it was in our best interest to allow foreign citizens to hold chairs in Italian universities, something that was impossible until the 1970s. Further, many of his writings show evidence to his scornful and lucid criticism of the contradictions inherent in present-day economic and social systems, expressed without mincing words, and often quite provocative. For example, in Dall'utopia all'alternativa, *he states that the aim of mathematical economics is to search for* "situations favourable for the quality of life of populations" *but instead* "the only questions that are raised are at the level of businesses, and have as their objectives not the best and least costly service for the consumer, but rather the maximum profit to the business". *He also decries the fact that* "every freedom, beginning with that of the press, is in fact only effective for those who have the means to distort it"; *to make his criticisms more effective he often makes up his own words, such as the terms* "bureauphrenic" *and* "juridiculous" *that we saw earlier, or* "deformation press" *(instead of information); he calls money* "the devil's shit", *an allusion to the proverb of Trieste* "the devil shits on the largest pile".

His commitment to problems of the environment was noteworthy and farsighted, leading him to remark that "to the traditional commandments there needs to be added – in recognition of the danger of future damage – 'thou shalt not pollute', 'thou shalt not waste', 'thou shalt not destroy', 'thou shalt not alter the ecological equilibrium'".

In the late 1970s de Finetti became a supporter of Marco Pannella's Radical Party and accepted the role of the managing director of the newspaper Notizie radicali. This also led to his being arrested for having published an article in that newspaper in defence of conscientious objectors. His entrance into the prison of Regina Coeli caused an uproar, but he was freed even before he entered his cell, the arrest warrant having been immediately revoked.

Andrey Nikolaevich Kolmogorov

The Foundations of Probability. And More...

Guido Boffetta and Angelo Vulpiani

April 25, 2003, marked the centennial of the birth of Andrey Nikolaevich Kolmogorov, probably the greatest Soviet mathematician of the twentieth century. He was born out of wedlock – his surname is that of his maternal grandfather – and his mother died in childbirth, so he was raised by a maternal aunt who instilled in him a strong sense of personal responsibility and intellectual independence. After finishing school, he worked for a time as a railway conductor before entering Moscow State University in 1920. These were hard years in the fledgling USSR: when he learned that second-year students, in addition to a meagre stipend, also received an additional monthly ration of 16 kg of bread and 1 kg of lard, he immediately stood for the examinations to pass to second year.

Andrey Nikolaevich had shown himself to be precocious from the start. By the time he earned his degree in mathematics, in 1925, he already had several scientific publications to his credit, including a fundamental work of 1922 in which he constructed a function that could be integrated with a Fourier series that diverges almost everywhere, which led to his becoming known worldwide.

By the time he finished graduate school, at 26-years old, he had already laid the foundations of modern probability theory. 1933 saw the publication of his monograph *Grundbegriffe der Wahrscheinlichkeitsrechnung* (Foundations of the Theory of Probability), which is probably the most important work of the first phase of Kolmogorov's career. Here he set out the theory of probability on an axiomatic basis, thus surmounting the historic dispute between those who adhered to frequency interpretations and those who supported subjective probability. It is no exaggeration to say that Kolmogorov's *Grundbegriffe* was as significant for the calculus of probability as Euclid's *Elements* were for geometry.

After earning his doctorate, Andrey Nikolaevich took off for a vacation boating and camping along the Volga (with equipment made available to him by the Soviet Society for Tourism and Proletariat Excursions), passing through the Caucasus Mountains to the Caspian Sea, in the company of his friend, the mathematician Pavel Sergeevich Aleksandrov. They spent months on rivers, lakes and in the mountains, but they also worked on Markov processes. The friendship between the two would turn out to be lifelong. In 1931 Kolmogorov was named professor at

C. Bartocci et al. (eds.), *Mathematical Lives*,
DOI 10.1007/978-3-642-13606-1_17, © Springer-Verlag Berlin Heidelberg 2011

Moscow State University. This is where he carried out all of his scientific work, except for brief stints in France and Germany.

Andrey Nikolaevich Kolmogorov

His scientific work is so vast and diverse that it is almost impossible to sum it up in the space of a few pages. His research in mathematics went from logic to stochastic processes, to analysis and to the theory of automata. In his contributions – even in the briefest of these – Kolmogorov never dealt with isolated problems, but rather shed light on fundamental aspects and new fields of research. It was precisely because of the broad panorama of his research, various scholars – even experts in mathematics – are familiar with only particular aspects of his multifaceted activity. His student V. I. Arnold recalled, somewhat ironically, "In 1965 Fréchet said to me, 'Kolmogorov, isn't he the brilliant young man who constructed a function that could be integrated with Fourier series that were divergent almost everywhere?'. All of the later contributions of Andrey Nikolaevich – in theory of probability, topology, functional analysis, turbulence theory, theory of dynamic systems – were of lesser value in Fréchet's eyes".

Here, then, we will limit ourselves to touching on some aspects of Kolmogorov's influence on modern (and not only strictly mathematical) areas of research: chaos, turbulence, complexity, and the mathematical description of biological and chemical phenomena.

Kolmogorov's interest in the theory of probability was not restricted to the merely technical and formal levels: he would in fact lay the foundations for the theory of stochastic processes, leading him, in the 1940s and 1950s, to deal with various problems in physics and biology. In many of these works his contribution went so far as to revolutionise the way the problem is seen. One example of this is found in his studies of turbulence, where his works are still, 60 years later, one of the few fixed points of reference for our comprehension; even today this complex phenomenon is not entirely understood.

The problem of fully developed turbulence (that is, of the irregular motion of fluids at large *Reynolds numbers*) gives an idea of Kolmogorov's extraordinary versatility. He alternates a formally mathematical study of the problem with a statistical analysis of the experimental data regarding atmospheric turbulence. In his preface to a recent book, Ya. G. Sinai writes, "When Kolmogorov was about 80-years old, I asked him about his discovery of the scaling law. He gave me an astonishing answer, saying that he had studied the results of experimental measurements for about a half a year".

His theoretical description of fully developed turbulence was of the broadest generality: the introduction of the concept of invariance of scale is in fact the root of the method of the renormalisation group developed in the 1970s. A second fundamental contribution to turbulence, in the early 1970s, stimulated by precise experimental measurements and by the observations of the great theoretical physicist Lev D. Landau regarding intermittent fluctuations of dissipated energy, was the starting point for studies (still ongoing) on anomalous fluctuations at a small scale. Kolmogorov's *log-normal theory* of turbulence, even though today partly superseded, forms the basis for intermittent stochastic processes (multi-affine or multifractal) which are used today in applications ranging from economics to geophysics.

At the end of the 1930s Kolmogorov – in collaboration with Petrovsky and Piskunov – studied the spatial evolution of biological species, introducing a system of mathematical equations that became the starting point for modern studies of reaction-diffusion systems. From this pioneering study was born the sector of systems of partial differential equations of reaction-diffusion, which finds applications that range from the spread of epidemics to the evolution of complex chemical processes such as ozone equilibrium and combustion. While dealing with biological problems in Stalinist Soviet Union, Kolmogorov courageously went up against the powerful academic Lysenko (who undertook a vehement campaign against Mendel's genetics, which in his opinion did not conform to dialectic materialism): this was a dispute that several eminent Soviet biologists would pay dearly for.

Another modern field of research indissolubly tied to the name of Kolmogorov is that of theory of nearly-integrable Hamiltonian systems. Already at the end of the nineteenth century Poincaré, studying the so-called three body problem in celestial mechanics (that is, the motion of the planets around the sun or the sun–earth–moon systems) had shown that when a small perturbation is added to an integrable Hamiltonian system, in general the motion is no longer integrable, and its behaviour can be chaotic. In dealing with this problem, Kolmogorov formulated a fundamental theory, later refined by V. I. Arnold and J. Moser (the KAM theory), which led to a reconsideration of some well-established (but erroneous) convictions (for example, on the generic ergodicity of Hamiltonian systems). In spite of the fact that nontrivial prime integrals do not exist, if the turbulence is small – in suitable hypotheses –

there are invariant tori (which are deformations of those that are unperturbed) in a set whose dimension tends to 1 in the integrable limit. KAM theory is now a flourishing area of mathematical physics, with applications ranging from celestial mechanics, to instability in planetary physics and to the foundations of statistical mechanics.

Other contributions by Kolmogorov to the development of twentieth-century science regard information theory and the definition of complexity.

Kolmogorov was one of the few mathematicians who immediately grasped the conceptual and not only practical relevance of Shannon's theory: "I recall that as far back as the ICM in Amsterdam (1954) my American colleagues, specialists in probability theory, regarded my interest in Shannon's work as somewhat exaggerated, since this was more technology than mathematics". Today such opinions carry no weight at all. The mathematical systemisation of information theory took place in the second half of the 1950s, primarily the work of Khinchin, Gel'fand and Yaglom, as well as Kolmogorov himself. Of particular significance is the use of concepts of information theory in the context of dynamic systems, with the introduction of what is known today as *Kolmogorov-Sinai entropy*. This intrinsic quantity (that is, a quantity that is independent of the variable used) measures the information generated per unit of time in chaotic systems.

In 1965 Kolmogorov proposed a measure (unambiguous and mathematically well founded) for the complexity of an object (for example, a string of bits) as the length of the shortest computer program required to reproduce the string. This topic, initially linked to a very particular context (an apparent "defect" of probability theory which assigns the same probability to whether the successive fair coin tosses will result in 0 or 1), has then developed, giving rise to a prolific area of research: algorithmic complexity. This sector has turned out to be extremely general and important due to its connections to chaos, Gödel's theorem and the application to problems of the most diverse natures, from linguistics, to the study of DNA sequences, to the analysis of financial trends.

These few pages are certainly not adequate to give a true picture of Kolmogorov either as a scientist or a man. As a great theorist, he was able to deal in a profound way with "practical" problems (such as turbulence and biological phenomena), opening new lines of mathematical research. Similarly, beginning with fundamental themes (such as complexity), he made contributions to the development of an area of research that today finds practical applications in computer science. His work is perhaps the best proof of the fact that fundamental science and applied science are not distinct; rather, we have one science and its application.

Kolmogorov was also extremely active in spreading notions of science, both as an author, writing more than a hundred entries for the *Great Soviet Encyclopedia*, and as a teacher. He was particularly interested in mathematics education for teenagers, an age when, in his opinion, the scholastic system had not yet succeeded in convincing them that science was useless. More than 60 of his students earned their doctorate degrees (including many who would become important scientists). He loved to spend at least a couple of days a week at his *dacha* in Kamarovka, near Moscow, discussing mathematics and competing with them in skiing and running: "Especially did we love swimming in the river just as it began to melt . . . I swam only short distances, but Aleksandrov swam much further".

In addition to mathematics, Kolmogorov was particularly interested in history, linguistics and literature (especially in the forms and structures of the poetry of Pushkin), and published articles in specialised journals.

His enthusiasm for all aspects of science led him (at almost 70) to take part in two oceanographic campaigns lasting several months (the Baltic, the Atlantic, the Panama Canal, the Pacific, and then back to Moscow on the Trans-Siberian). He died in Moscow in 1987. He had been the recipient of numerous prizes and honours, but above all, he left behind a scientific legacy that will survive for a long while.

Bourbaki

A Mathematician from Poldavia

Giorgio Bolondi

France, 1930s. In spite of the fact that many years had passed, France was still in shock over the horrendous massacre that bloodied Europe between 1914 and 1918. One million three hundred thousand dead, three million maimed and wounded, eight hundred thousand widows, almost a million orphans. This kind of tragedy couldn't help but affect all aspects of a nation's life. Of course, the handing down from generation to generation of mathematics (as of all sciences in general) was of course drastically disrupted as well by the great war. In the halls of the *Grandes Écoles* hung enormous plaques with endless lists of students and professors who had died in the trenches. An entire generation had been swept away: of 211 students enrolled in the École Normale in 1914, 107 died in the war.

As things stood in 1930, there was little left of the great mathematics of France of the beginning of the century – the mathematics of Poincaré, Lebesgue, Fatou – even the great names still living had lost the majority of their students. Teaching stagnated, and the textbooks then in use mostly dated from before the war.

Two young, brilliant university professors, just named to their chairs, expressed their dissatisfaction with the text used everywhere for teaching analysis, Edouard Goursat's *Cours d'Analyse*, which they criticised as by then obsolete, and above all, lacking in rigour. They decided to compile a new text, one that would be as rigorous as possible.

C. Bartocci et al. (eds.), *Mathematical Lives*,
DOI 10.1007/978-3-642-13606-1_18, © Springer-Verlag Berlin Heidelberg 2011

II General Bourbaki

Thus began the Bourbaki adventure, a revolution destined to leave its mark on twentieth-century mathematics; a revolution which, like all revolutions, soon aspired to expand beyond the French border and which, in the course of a few decades, would become "establishment".

The two young professors, André Weil and Henri Cartan, organised meetings in a restaurant on the Boulevard Saint-Michel in Paris with a group of former classmates at the *École Normal*, with the aim of compiling a new, collective text for analysis.

After the first meetings, it soon became evident that the work was going to require collaboration and plenty of time. The first Bourbaki "congress" took place during the summer holidays, in July 1935, in Besse-en-Chandesse, a small town about 50 km from Clermont-Ferrand. Participating were Henri Cartan, Claude Chevalley, Jean Dieudonné, Jean Delsarte, Szolem Mandelbrojt, René de Possel and André Weil (some name Charles Ehresmann as well). Almost right away it was decided to adopt the name of Nicholas Bourbaki as a pseudonym, and almost as immediately the founders began to delight in leaving false clues as to their true identity, wrapping the group's history in a cloak of anecdotes and mystery. Élie Cartan was also accomplice to this. When a paper was submitted for the *Comptes Rendus* of the Académie des Sciences in Paris, the author had to of course be named, and even more importantly, the paper and the author's biography had to be presented by member of the Académie. The presenter was Élie Cartan, Henri's father, and he convinced the members of the Académie of the importance of this unknown mathematician from "Poldavia" (the imaginary country where most of the group's farcical tales were set). It was partly due to this that Élie Cartan is considered Bourbaki's godfather.

The second congress, which was to have taken place in Spain, was instead held at Chevalley's mother's house in Chancais, because of the outbreak of the Spanish Civil War. It marked a turning point in the group's project. In order to write the text for analysis in a rigorous manner, it was first necessary to write a treatise which was a systematic treatment of the theorems and results preliminary to the treatment of all existing (and future) mathematical theories. Thus was born the project of the *Eléments de Mathématique,* which, as the title implies (in French, as in English, the word "mathematics" is used in the plural) presages an overarching treatment of the subject.

When all is said and done, this unifying treatment of mathematics was the great objective of the Bourbaki group. To construct the whole beginning from a common root, a root that had to be searched for in the structural hierarchy (algebraic, of orders, topological), starting from the most general and abstract and proceeding to an axiomatic explanation. Hilbert is often referred to as Bourbaki's spiritual father. To be sure, there was no lack of reference to exponents of the German school: Emmy Nöther, for example, but also the Dutch mathematician van der Waerden (and his way of structuring his *Moderne Algebra*). The 1947 article "L'architecture des mathématiques" signed by Bourbaki himself was the group's ideological manifesto. Jean-Pierre Kahane has written that the history of mathematics has shown that the unity of the discipline has to be looked for in the way its branches interweave, not in the unity of its roots; but the fact remains that the project undertaken by Bourbaki profoundly changed the way that mathematics was done (and written about).

The group's working method was truly collective: a theme was chosen, and lively discussions ensued about how the work was to be set up (minutes were taken by a "scribe", usually Dieudonné, and later Cartier). Someone was charged with laying out a first draft which was then sent to all members. At the next congress, it was decided what was to be done with the material, whether to accept it, rework it, or trash it. The Paris archives of the Bourbaki group contain thousands of pages of mathematics that have never been published.

One rule that was decided on during the second congress was that of an age limit: members agreed to leave the group completely when they reached 50 (in fact, Bourbaki himself published just one thing, in 1998, after his 50th birthday: the tenth chapter of the "Algébre Commutative", a book began in 1964).

Thanks partly to the age limit, and thus the continual turnover generation after generation, there were never more than a dozen members of the group at any given time. Between the rules established, the enthusiasm of the members, and great ambition, the first generation of Bourbaki accomplished the writing of a certain number of volumes of the *Eléments*, aimed at professional mathematicians rather than students. As the Bourbakis themselves said repeatedly, the *Eléments* had to be thought of as an encyclopaedia because, taken as a textbook, it was a disaster.

The mathematical merit of Bourbaki members (including some of the greatest mathematicians of the twentieth century: Laurent Schwartz, Jean-Pierre Serre, Alexander Grothendieck, Alain Connes, Jean-Christophe Yoccoz, all members of Bourbaki, and all Fields Medal laureates), and the added merit of their "collective" weight, was such that in the end the group influenced mathematics throughout the world. Bourbaki can in fact be considered the mathematician who had the greatest influence on the training, working methods, and writing style of the majority of today's mathematicians. It was Bourbaki who introduced the symbol ∅ to indicate an empty set. It was again Bourbaki who spoke for the first time of injective, surjective and bijective correspondences; of filters and ultrafilters, of germs, of separated spaces and paracompact spaces, and so forth. From the 1950s on, Bourbakism was the dominant formulation (or ideology, according to your point of view) of mathematics in many countries, especially (and obviously) in France. For instance, the late development of applied mathematics in France is usually attributed to Bourbaki's lack of interest in applications.

It was thanks above all to the work of Jean Dieudonné that there was an attempt to transform mathematics teaching in a Bourbakian sense. His famous cry of "Down with Euclid!" still today indicates the desire to do away with the teaching of geometry, of figures, and of intuition (although the reaction against this of many mathematicians, starting with René Thom, in support of the educational importance of geometric teaching, is equally famous). The introduction of set theory in schools was the offspring of Bourbaki. Legitimate offspring? Some have spoken of a degeneration, which has led many to believe that the essence of mathematics consists in the "vocabulary" and definitions. Others believe that the Bourbaki group is directly responsible for this "reform" of teaching, even though it then tried to "draw back its hand".

Starting at the end of the 1960s, with the dawn of a new generation, Bourbaki's actions were less cohesive. It became increasingly difficult to find shared objectives once the "founding" period of the group had passed. Figures such as Grothendieck left the group in a huff. Other mathematicians (great and not so great) learned to write like Bourbaki, but the gaps (mathematical logic, numerical analysis, combinatorics, mathematical physics, probability theory) became ever more evident. Further, Bourbaki found itself caught up in a neverending lawsuit with its publisher, and in the end its members could not agree on what direction to take. In fact, after 1983 no more publications appeared, except from the above mentioned tenth chapter of the "Algébre Commutative".

Thus we see how Bourbaki was born, and how he lived. But can we say he is still alive, seeing as how nothing has been published for more than 20 years?

Already in 1968 Jacques Roubaud, a mathematician and author who was a member of *Oulipo* (*Ouvroir de littérature potentielle*, which translates roughly as "workshop of potential literature"), had announced the death of Bourbaki, saying that he had been buried in the cemetery of random functions, alongside the Markov number and the Gödel number, and that the funeral mass had been said in the church of Our Lady of Universal Problems.

According to Pierre Cartier, who had been a pillar of the Bourbaki group for about 30 years, the group had reached its own physiological limit. Even the mathematics project had exhausted its reason for being, without having even achieved its own aims. It had been a dogmatic vision of mathematics, child of the century of ideology (to quote Cartier), self referential (there were no citations except for internal references): a great mathematical enterprise, the greatest of the twentieth century, but perhaps not great enough to become – as its members had aspired – the new Euclid, the fundamental textbook for the coming millennium. There were other surprises, other revolutions in store for mathematics, which not even Bourbaki could have imagined (Fig. 1).

Les familles Cantor, Hilbert, Nœther ;
Les familles Cartan, Chevalley, Dieudonné, Weil ;
Les familles Bruhat, Dixmier, Godement, Samuel, Schwartz ;
Les familles Cartier, Grothendieck, Malgrange, Serre ;
Les familles Demazure, Douady, Giraud, Verdier ;
Les familles Filtrantes à Droite et les Epimorphismes Sricts ;
Mesdemoiselles Adèle et Idèle,

ont la douleur de vous faire part du décès de Monsieur

Nicolas BOURBAKI

Leur père, frère, fils, petit-fils, arrière-petit-fils et petit-cousin respectivement.

Pieusement décédé le 11 Novembre 1918 (jour anniversaire de la Victoire), en son domicile de Nancago.

L'inhumation aura lieu le Samedi 23 Novembre 1968, à 15 heures, au cimetière des Fonctions Aléatoires (Métros Markov et Gödel).

On se réunira devant le bar Aux Produits Directs, carrefour des Résolutions Projectives (anciennement place Koszul).

Selon le vœu du défunt, une messe sera célébrée en l'église Notre-Dame-des-Problèmes-Universels par S.E. le Cardinal Aleph 1, en présence des représentants de toutes les classes d'équivalence et des corps (algébriquement clos) constitués. Une minute de silence sera observée par les élèves des Ecoles Normales Supérieures et des classes de Chern.

Ni fleurs ni wreath products.

« Car Dieu est le compactifié d'Alexandrov de l'univers. » Groth. IV. 22.

Fig. 1 Bourbaki's "obituary"

Books Published by Bourbaki

Here is the list of volumes published up to the present in the *Eléments de Mathé-matique*:

1. Set theory (*Théorie des ensembles*)
2. Algebra (*Algèbre*)

3. Topology (*Topologie générale*)
4. Functions of one real variable (*Fonctions d'une variable réelle*)
5. Topological vector spaces (*Espaces vectoriels topologiques*)
6. Integration (*Intégration*)
7. Commutative algebra (*Algèbre commutative*)
8. Differential and analytic manifolds (*Variétés différentielles et analytiques*)
9. Lie groups (*Groupes et algèbres de Lie*)
10. Spectral theory (*Théories spectrales*)

The Generations

Here are the first four generations of Bourbaki, according to Pierre Cartier:

Founding fathers: Henri Cartan, Claude Chevalley, Jean Delsarte, Jean Dieudonné, André Weil (who were soon joined by Jean Coulomb, Charles Ehresmann, Szolem Mandelbrojt and René de Possel).

The war generation: Jacques Dixmier, Samuel Eilenberg, Roger Godement, Jean-Louis Koszul, Pierre Sameul, Laurent Schwartz, Jean-Pierre Serre.

The third generation: Armand Borel, François Bruhat, Pierre Cartier, Alexander Grothendieck, Serge Lang, John Tate.

The fourth generation: Michael Atiyah, Louis Boutet de Monvel, Michel Demazure, Adrien Demazure, Bernard Malgrange, Jean-Louis Verdier, . . .

A question: Was there ever a female Bourbaki?

The "Real" Bourbaki

A French general of Greek origin, Charles-Denis Sauter Bourbaki lived from 1816 to 1897. He took part in the Crimean war (where he led his troops to victory in several battles) and in the French–Prussian war of 1870 (where he was defeated). His name began to be used in jokes in the Ecole Normal as early as 1880, when a student gave the name Bourbaki while posing as an official who had come to inspect the school.

According to Andre Weil's autobiography, in 1948 a Greek diplomat named Nicolaides Bourbaki introduced himself to Henri Cartan as a member of General Charles-Denis's family. There had never been a mathematician in that family, but from then on this one was invited to all the closing dinners of the Bourbaki congresses. A true story?

The Founding Fathers

Jean Dieudonne, the scribe of the first generation of Bourbakis and a strong, domineering personality, was born in Lille in 1906. He entered the École Normale in 1924, and in 1928 he went to Princeton on a scholarship. Even before completing his thesis, which he wrote when he was 25, he had been given the opportunity to work around the world with mathematicians of the calibre of Bierberbach and Pólya. In the 1960s, in a surprise move, he left the Paris Institut des Hautes Etudes Scientifiques (IHES) and transferred to the newly established Faculty of Sciences in Nice, as yet completely without mathematicians. This had the effect of shifting the barycentre of French mathematics towards the Cote d'Azur. The last years of his life were dedicated to the history of mathematics (as seen by the Bourbakis), and he published one of the most successful mathematics books in recent years, *Pour l'honneur de l'esprit humain*. He died in 1992. His work mainly concerned topological vector space, topology (the notion of paracompact space is due to him), and the theory of Lie groups.

Jean Delsarte, of all the founding fathers perhaps the least well known, was born in 1903. He first came into contact with Cartan and Weil (at that time in Strasbourg) in the 1930s. Delsarte spent a good part of his professional life in Nancy, which thanks to him became an important centre for mathematical research (this is where Dieudonne, Schwartz and Godement "grew up"). Of a very fragile constitution, the events of 1968 took a heavy toll on him, and he died of a heart attack in November of that year.

Claude Chevalley, the youngest of the founders, was born in 1909 in Johannesburg, the son of a diplomat. His whole life was a weave of his interest in philosophy, of which he made a way of life, with his interest in mathematics. He was in agreement with the epistemology of Meyerson, and was connected to Arnaud Dandieu in activities of the 1930s movement called "Ordre nouveau", a nonconformist movement that verged on anarchism. Also in the 1960s he founded, together with Alexander Grothendieck and Roger Godement, also Bourbakis, the ecological movement "Survivre et vivre". His most important mathematical contributions were in number theory (the class field theory), and in algebraic geometry. It was Chevalley who introduced the terms "injective" and "surjective". Some of his writings have become classics. He died in 1984.

André Weil, born in 1906, went to Rome to study with Vito Volterra when only 19-years old. He then went to Berlin, Göttingen and Stockholm (coming into contact with Emmy Nother and Gösta Mittag-Leffler along the way). At 22 he completed his doctoral thesis, and in 1937 he married Eveline, former wife of Rene de Possel, (another member of the first generation of Bourbakis). His personal actions during the Second World War were the subject of controversy: he was accused first of desertion (a choice influenced by his ties to Oriental philosophy), and then of spying for the Soviet Union (which would be more difficult to connect to his dharma). Saved from execution thanks to Rolf Nevanlinna, he still had to face the resentment and incomprehension of colleagues –notably, Jean Leray – who had experienced the drama of war and prison first-hand. Weil was able to take refuge in the United States, thanks to a program established by the Rockefeller Foundation to provide safe haven for French scientists. After the war, Leray's opposition kept him from going back to the Sorbonne, and he remained in Princeton until his death on 6 August 1998.

His works in number theory and algebraic geometry are fundamental and foundational. From results regarding the Riemann hypothesis to studies of the arithmetic of elliptical curves, his contributions made new in-roads in mathematics. The Shimura–Taniyama–Weil conjecture was the crucial point in Andrew Wiles's proof of Fermat's Last Theorem. André's sister, Simone Weil, took part in the first meetings of the Bourbaki group and is the only woman to appear in the group's first photograph. She was awarded the Wolf Prize in 1979 (along with Jean Leray!).

Henri Cartan was born in 1904 in Nancy, and followed in the footsteps of his father, mathematician Élie Cartan, one of the founders of modem differential geometry. Like all founders of the Bourbaki group, he studied at the École Normale, and after having started his career in Caen, Lille and Strasbourg, he soon returned there, staying until 1965. In 1935 he married Nicole Weiss, daughter of physicist Pierre Weiss. In 1980 he shared the Wolf Prize with Andrey Kolmogorov. His works on functions of several complex variables are fundamental: he introduced the notion of "sheaf" into the geometry of analytical space (a notion which had been created in algebraic topology by Jean Leray). Also owing to Henri Cartan is the concept of "filter" in topology.

He was a pioneer in the efforts to harmonise mathematical studies in Europe. His objective was to facilitate programs for student exchange between countries. He also worked intensely to mend the fissure created by the Second World War (similar to that created by the first world war) between German and English mathematicians. His tolerance is remarkable in light of the fact that his own brother, Louis Cartan, was executed by the Nazis in 1943 for being a member of the resistance.

Writing and Mathematics in the Work of Raymond Queneau

Alessandra Ferraro

Just a few years ago was celebrated the centennial of the birth of Raymond Queneau, who was born in Le Havre in 1903 and who became, after earning a degree in philosophy at the *Sorbonne*, a writer and essayist as well as general secretary of the prestigious Paris publishing house *Gallimard*, a position that he held until his death in 1976. Author of the amusing *Zazie dans le métro*, a novel written using a phonetic writing to illustrate how the French language had changed, his works were translated into Italian, with translations done by none other than Italo Calvino for *Petite cosmogonie portative*, and Umberto Eco for *Exercices de style*, both published by Einaudi. An unflagging reader with a vast store of knowledge, Queneau came into contact with the principle literary and cultural movements in the Paris scene, from Breton's surrealism to Sartre's existentialism, without ever adhering to any of them unconditionally. This intellectual independence gave rise to a multifaceted output, including novels, poetry, short stories and screenplays, all original and unclassifiable. Only recently have critics begun to discover the profound unity of the poetic concept that underlies them all, from the first novel, *Le Chiendent* of 1933, to the works composed in the context of Oulipo (derived from *Ouvroir de littérature potentielle*), the workshop for potential literature, a working group that Queneau founded in 1960 with his friend, the mathematician Françoise Le Lionnais. But that this was more than an amusement can be understood from reading his diary from his years in high school and at university. At 17 he wrote, "I went with Leroux to the Museum. I am furiously studying mathematics".[1] We find traces of this passion in Roland Travy, the main character in the autobiographical *Odile* (1937). In 1921 the young Queneau remarked again, "Leafing through my papers I realise that at 13 I discovered the algebra of logic".[2] A voracious reader of scientific works and a follower of the latest scientific theories since he was a teenager, he enrolled in the Faculty of Philosophy, where he attended courses in logic and mathematics, with particular attention to philosophical literature related

---

[1] Raymond Queneau, *Journaux 1914-1965*, Anne Isabelle Queneau, ed. Gallimard, Paris, 1996, p. 51.
[2] Ibid., p. 73.

C. Bartocci et al. (eds.), *Mathematical Lives*,
DOI 10.1007/978-3-642-13606-1_19, © Springer-Verlag Berlin Heidelberg 2011

to mathematics. Already in 1919, at 16, he remarked in his diary on the importance of Einstein's discovery: "A certain Einstein seems to have made sensational discoveries (see the newspaper of the 10th). Rutherford is supposed to have decomposed nitrogen into hydrogen. Tried to read Proust: soporific".[3] In the years that followed he pored over Cantor's *Contributions to the Founding of the Theory of Transfinite Numbers*, Einstein's *Theory of Relativity*, Poincaré's *La science et l'hypothèse*, to name only a few of the many scientific works he read. During those years he also attended the courses of Pierre-Lèon Boutroux, professor of history of science at the *Collège de France*. This interest, probably coupled with his failure to pass some exams in philosophy, led him to enrol in the Faculty of Sciences with a major in mathematics. But his academic performance was disappointing. This is how Raymond Queneau commented many years later on the collapse of his university career in mathematics: "Where I made my mistake was in believing that I could fill the gaps. . . . I became perfectly aware when I enrolled in my first year of mathematics. After failing two or three exams I understood that I would never pass. For example, for me mechanics was opaque. And so were the conics (the delight, the *non plus ultra* of mathematics for specialists. . .)".[4]

However, Queneau continued his whole life through to nourish a passion for mathematics. This expressed itself not only through specific readings and constant practice, but also through his attending the seminars of the most important mathematicians then working in Paris. In 1948 he joined the *Société mathématique de France* and in 1963 the American Mathematical Society. From that year he also attended the seminars in operation research and graphs theory, and consulted with A. Kaufmann and R. Faure on their book *Invitation à la recherche opérationnelle*. From his diary we know that in the 1950s he attended the meetings of Bourbaki and that he regularly went to dinner with Georg Kreisel, with whom he discussed the main mathematical innovations. This was not just pure curiosity, because he collaborated on the *Elements de logique mathématique* published by Kreisel and J.-L. Krivine in 1967.[5]

But above all, evidence of his research in the field is given by the presentation of the results of his work in number theory at the *Académie des sciences* in Paris in April 1968.[6] This was followed by a work on s-additives series, with commentary by another great mathematician, Gian Carlo Rota, published in the *Journal of Combinatory Theory*.[7]

---

[3]Ibid., p. 44.

[4]Anne Isabelle Queneau, ed. *Album Raymond Queneau*, Gallimard, Paris, 2002, pp. 43–44.

[5]Georg Kreisel, Jean Louis Krivine, *Éléments de logique mathématique, théorie des modèles*, Dunod, Paris, 1967 (Monographies de la Société mathématique de France 3).

[6]As is customary, the report on the results was given by a mathematician, André Lichnerowicz, a member of the *Académie des sciences*, followed by a publication in the *Bulletin de l'Académie des sciences*.

[7]Raymond Queneau, "Sur les suites s-additives", *Journal of Combinatory Theory*, 12, 1972, pp. 331–371.

This is a significant result for someone who professed to approach mathematics as a dilettante, with occasional peripheral forays into the subject, as indicated by the title of the 1963 book *Bords* (*Borders*),[8] a collection of Queneau's essays on people such as Hilbert, Bourbaki and Léopold Hugo, nephew of Victor Hugo, a geometer but mad. This is similar to what had been done in other branches of the sciences – philosophy, history, medicine – as can be clearly seen in the summaries of the literature that can be found in the Italian collection of Queneau's essays edited by Italo Calvino entitled *Segni, cifre e lettere*,[9] and which follow along the same lines as the monumental *Encyclopédie de la Pléiade* which Queneau worked on for several years. Beyond the poetic works such as *Petite cosmogonie portative*, which is a kind of hymn in praise of science, or the *Le chant du styrène*, which tells about the adventure in chemistry when plastic was created, this mastery of scientific subjects, which was genuinely exceptional for a literary man in the modern age, underlie – though often in ways that are hidden, encrypted or dissimulated – the majority of Queneau's works.

Beyond the judgments of the critics regarding the merits of individual novels or collections of poetry, it seems to us that the element that renders Queneau's work unique in the contemporary panorama in France and beyond lies in his having constructed an "amalgam" of science and literature without its having weighed down the narration, which remains light and amusing, as in *Les fleurs bleues* or *Zazie dans le métro*. Queneau was able to experiment with new literary structures and innovative linguistic forms without leaving traces in the writing of the kind of uncertainties that often accompany creative efforts when they depart from the beaten path.

The image chosen by Raymond Queneau to characterise his poetry is that of an onion: every ring of the onion corresponds to a level of possible interpretation of the work, each as valid as the next. And the mathematical ring was a constant in the construction of his works from the beginning: in theoretical writings and in interviews the writer underlined how each novel was based on complicated calculations and rigorous constructions: "Even for linear novels ... I've always forced myself to follow certain rules that had no other reason for being than to satisfy my taste for numbers or other strictly personal fancies".[10] Referring to his first novels, Queneau says, "I found it intolerable to leave the number of chapters in these novels to chance. For this reason *The Bark Tree* is composed of 91 (7 × 13) sections, 91 being the sum of the first thirteen numbers and its own "sum" being 1; it is thus the number of the death of living things and of their return to existence, a return that at the time I conceived as the irremediable perpetuity of hopeless suffering".[11]

---

[8]Raymond Queneau, *Bords – Mathématiciens Précurseurs Encyclopédistes*, Hermann, Paris, 1963.

[9]Raymond Queneau, *Segni, cifre e lettere e altri saggi*, Italo Calvino, ed., Einaudi, Torino, 1981.

[10]Raymond Queneau, "Conversation avec Georges Ribemont-Dessaignes", in *Bâtons, chiffres et lettres*, Gallimard, Paris, 1965, p. 42. English translation from *Letters, Numbers, Forms: Essays 1928-70*, trans. Jordan Stump, University of Illinois Press, 2007, p. 177.

[11]Raymond Queneau, "Technique du roman", in "Conversation avec Georges Ribemont-Dessaignes", in *Bâtons, chiffres et lettres*, Gallimard, Paris, 1965, p. 29. English translation

Raymond Queneau

However, the work of Queneau that is most obviously influenced by mathematics is an essay written in 1942 entitled *"Brouillon projet d'un atteinte à une science absolute de l'histoire"* (Draft of a project to attempt to arrive at an absolute science of history), published in incomplete form in 1966 with the title *"Une histoire modèle"* (A Model History). Here Queneau bases himself on Vito Volterra's biomathematics[12] in the attempt to formulate a model that can be applied to the evolution of human history.

Among the works of creative literature we can cite the 1947 *Exercices de style*, which recounts the same banal incident in 99 different ways. The publication of *Cent mille millards de poèmes* dates to 1961, a singular book composed of ten sonnets, where each of their respective 14 lines of verse, all of the same rhyme scheme and the same syntactic construction, appears on a separate strip of paper. The verses can therefore be combined so as to produce, as the title suggests, a hundred thousand billion poems.

While we certainly find the presence of a certain "arithmania" forming the basis of these works, as Queneau himself acknowledged, in reality this serves a

---

from *Letters, Numbers, Forms: Essays 1928-70*, trans. Jordan Stump, University of Illinois Press, 2007, p. 27.

[12]Vito Volterra, *Leçons sur la théorie mathématique de la lutte pour la vie*, Paris, Cahiers Scientifiques, 1931.

precise poetic intention which favours the formal aesthetic. The first literary experience related to the group embracing surrealism, which Queneau began to take part in starting in 1924, only to remove himself stormily 5 years later along with some other well-known deserters, had led him to develop a concept of literature that was in opposition to the surrealistic one based on a romantic matrix. Queneau thus refuted the idea that, in creating, the poet was guided by the unconscious, by automatism, by inspiration: for him, the authentic creator is the one who consciously imposes rules. As is clear, these considerations, which date back to the 1930s, constituted the prelude to the program of the Oulipo, which aimed at offering writers artificial literary structures in order to foster the task of creating.

From Exercises in Style by Raymond Queneau

Original Version

In the S bus, in the rush hour. A chap of about 26, felt hat with a cord instead of a ribbon, neck too long, as if someone's been having a tug of war with it. People getting off. The chap in question gets annoyed with one of the men standing next to him. He accuses him of jostling him anytime anyone goes past. A snivelling tone which is meant to be aggressive. When he sees a vacant seat he throws himself onto it.

Numerical Version

In a bus of the S-line, 10 m long, 3 m wide, 6 m high, at 3 km 600 m from its starting point, loaded with 48 people, at 12.17 p.m., a person of the masculine sex aged 27 years 3 months and 8 days, 1 m 72 cm tall and weighing 65 kg and wearing a hat 35 cm in height round the crown of which was a ribbon 60 cm long, interpellated a man 48 years 4 months and 3 days, 1 m 68 cm tall and weighing 77 kg, by means of 14 words whose enunciation lasted 5 s and which alluded to some voluntary displacements of about 15–20 mm. Then he went and sat down about 1 m 10 cm away.

Geometrical Version

In a parallelepiped rectangle moving along a straight line of equation $84x + S = y$, ovoid A wearing a spherical calotte encircled by two sinusoidal waves of length

$1 > n$ immediately below its crowning hemisphere manifests a point of contact with ovoid B. Prove that this point of contact is a cusp.

If ovoid A meets a similar ovoid C, then the point of contact is a disc of radius $R > 1$. Determine the height h of the point of contact in relation to the vertical axis of ovoid A.

Set Theory Version

In bus S, consider the set of seated passengers A, and the set of standing passengers B. At a certain stop, is the set P of people waiting. Let C be the set of passengers boarding; it is a subset of P, and is itself the union of set C' of passengers who remain on the platform and set C'' of the passengers who are going to sit down. Demonstrate that set C'' is empty.

Let Z be the set of the rude passengers, and $\{z\}$ the intersection of Z and C', containing a single member. As a consequence of the surjection of z's feet on those of y (an arbitrary but distinct member of Z), it is possible to determine a set M containing words spoken by z. Since set C'' has now become non-empty, prove that it contains only z...

John F. Nash Jr.

The Myth of Icarus

Roberto Lucchetti

John F. Nash Jr. was probably one of the most brilliant mathematicians of the twentieth century. His results are unanimously considered to be of the highest calibre, and have produced solutions to problems that were classified among the most difficult. His activity as a researcher lasted for just a short time, less than 10 years, before, as we will see, being forced to a halt. In this chapter, we first want to recount some details about Nash's life in a simple and clear way. The source for this is Nash himself, that is, his autobiography written on the occasion of the awards ceremony for the Nobel Prize in economics, which he won together with Harsanyi and Selten in 1994. Then I'll make some comments about the character of this genius, which has fascinated me since I read Sylvia Nasar's biography of Nash (*A Beautiful Mind: The Life of Mathematical Genius and Nobel Laureate John Nash*, 1998). Finally, I'll try to explain in simple language what game theory is, and will look at some of its first fundamental results, above all those in connection to von Neumann and Nash.

Nash's Life

Life "legally" began for John F. Nash Jr. on 13 June 1928 in Bluefield, West Virginia. His father was an electrical engineer, originally from Texas, a veteran of the First World War, who moved to Virginia to work for a local electric company. His mother was born in Bluefield, and was an English teacher who sometimes taught Latin as well. When Nash was 2-years old, his sister Marta was born. Nash had been an avid reader since he was a very young boy, and he felt pretty isolated in that community, which was mostly made up of businessmen, lawyers and salesmen. After finishing high school, he entered Carnegie Mellon University (then known as Carnegie Tech.) in Pittsburgh, where he majored in chemical engineering. He very soon left that, however, saying that the courses were too regimented, and turned instead to chemistry. But he came to discover that here too quantitative analysis was required, and that creative thought counted for less than certain skills in laboratory work. In the meantime, the math department managed to convince him that sometimes even a mathematician could make a good career. He switched majors again, and his undeniable talent became immediately evident, so much so that he graduated simultaneously with a bachelor's and a master's in mathematics. The letter of recommendation that he took away with him said unequivocally, "This man is a genius". This earned him several invitations for doctoral studies, in particular from Harvard and Princeton. In the end, after some hesitation, he settled on Princeton, which not only made him a more generous offer, but was closer to home. After some initial uncertainty about what to focus his research on, Nash earned his doctorate with a thesis on game theory, which brought him immediate fame, at least in academic circles. In the autobiography I mentioned earlier Nash tells of his position at MIT, the Massachusetts Institute of Technology, of the courses he taught at Harvard, and of his academic work. But there are other important aspects of his life that he does not mention: his work for the Rand Corporation, a small but active think-tank located in California. Under contract to the US government and the Marine to hire the best minds in game theory, Rand snatched up Nash, the most promising young mathematician of his generation. But his relationship with Rand was stormy, and by that time Nash was more interested in so-called pure mathematics than in game theory. He was let go after a few years. Rand also accused him of immoral behaviour following an incident that has never been entirely made clear, and this led to suspicions that he was homosexual. He also had a passing relationship with a woman who bore him a child, and though he never recognised him as his son, he continued to see the boy on and off and is still close to him today. In 1956 he married Alicia, a former student of his at MIT. In 1959 he resigned from MIT, because the mental illness that he had suffered from for some time became too obvious to hide. Nash suffers from paranoid schizophrenia, a mental illness that makes it impossible to work, and almost impossible to have any social life. There is no sense dwelling on the next 30 years, when John spent time in mental hospitals and clinics against his will, interspersed with periods of greater serenity, during which he travelled and even did some writing. What matters is that in the early 1990s, Nash, who had been called "the phantom", began to attend some seminars, to write some letters, to talk to colleagues. His long-time friends gradually tried to help him with his reintegration into society, and worked to see that he was given awards as publicly visible as the Nobel Prize, with a dual intent: on the one hand, it was thought that an award he had always aspired to would help him recover; on the other, more mundane, the award was accompanied by a handsome check, and it shouldn't be forgotten that for 30 years Nash had not had a job, much less a salary. The rest is recent history: first the Nobel, then articles about Nash, followed by Nasar's biography and the film based on Nash's story, "A Beautiful Mind" starring Russell Crowe, a somewhat romanticised version of his life that made Nash internationally famous. Today he lives in Princeton and from time to time travels, with frequent visits to Italy.

John F. Nash

The Myth of Icarus

This section takes its name from something that Sylvia Nasar said at the literary festival called *Festivaletteratura* in Mantua in September 2002, during a masterly talk that she gave on the occasion. The myth of Icarus is of course a daring flight up towards the light followed by a ruinous fall. Nash's life adds a new twist on the myth: after an overpowering ascent, and an apparently irremediable fall, there was a return to a normal life. We mentioned that at only 20-years-old Nash's doctoral thesis made him instantly famous, and not without reason, seeing as how more than 40 years later the results earned him the highest honour a scientist can hope for. And yet his restless mind led him to work on other problems. He wanted to grapple with more abstract ideas, and was obsessed with the idea of solving complicated questions, so complicated that their solutions would bring him global acclaim (at least among mathematicians). He was not as interested in specialising in one theory, even a very fashionable one, as he was in attacking problems that were deemed to be impossible, or almost. He even went around asking experts in the field for suggestions as to what problems to look into, and insisted on knowing whether or not they were sufficiently difficult. In a short time he solved three problems that earned him the lasting respect and admiration of the mathematics community. He was famous even outside that community, and a magazine once put his picture on the cover as one of the most promising scientists of his generation. He married a beautiful and very intelligent girl, one of the three (!) female students at MIT at the time. Like everyone, Nash too had periods that were dark or frustrating. He was very disappointed by the fact that one of his most prestigious results had in fact already been proven, although in a slightly less general way, and above all using different techniques, by Ennio De Giorgi, a great mathematician who will be discussed in a separate chapter. Earlier I mentioned his difficult relationship with the Rand Corporation, and the son that he never recognised. It is not only difficult but futile to determine exactly what brought on his illness, an illness that would have exploded one way or another. Paranoid schizophrenia is still not well understood even today. In Nash's case – but this is probably quite common, especially when schizophrenia is accompanied by paranoia – one symptom of the illness manifests itself when the person suffering from it begins to be obsessed with things that to everyone else seem obvious or unimportant: thus Nash saw conspiracies everywhere; he found messages where there were none, such as in random numbers on the pages of newspapers, messages that he claimed had come from strange and mysterious beings; he heard voices. Above all, and this is the paranoid aspect of the illness, he suffered from delusions of grandeur: he sometimes claimed to be the emperor of the universe. This is why when he was in Europe he caused a turmoil in various embassies, because he insisted on giving back his US passport. Other times he believed he was God's left foot. All of this went on, as he said, for about 30 years. Then something changed. Why, and more especially, how did he undergo this change? Reading his autobiography is illuminating. He began by deciding *intellectually* to push away some of the delusions that were basic to his way of thinking. And then he goes on to say, "However this is not entirely a matter of joy as if someone returned from physical disability to good physical health. One aspect of this is that rationality of thought imposes a limit on a person's concept of his relation to the cosmos". This gives the impression of someone would say who misses a significant part of the emotions, if not the thoughts, that he had had during his illness. It is certainly not like the story recently told by Marguerite Sechehaye in *Autobiography of a Schizophrenic Girl*, in which the main character, who had believed herself to be the queen of the Andes, was wonderfully happy to have returned to normalcy, and looked back with horror at the periods when she was ill. Nash insisted, "For example, a non-Zoroastrian could think of Zarathustra as simply a madman who led millions of naive followers to adopt a cult of ritual fire worship. But without his "madness" Zarathustra would necessarily have been only another of the millions or billions of human individuals who have lived and then been forgotten". This, written on the solemn occasion of the award of a Nobel! It is clear that Nash has always been obsessed with the idea of leaving something behind, just as it is clear that his hallucinations were tied to this dream, and that giving up the delusions also meant giving up the dream.

On 19 March 2003, the Federico II University in Naples awarded Nash an honorary degree, his first in economics, as strange as it may seem, seeing as how he had already won a Nobel in this field. On that occasion, he agreed to answer some questions. I was struck by two of his answers in particular:

> They say you are a genius. What do you think?
> It's difficult to talk about this.[1] If you ask someone who might be a genius if he really is one, you'll put him on the spot. If you were to ask Mozart why he is a genius and Hayden isn't, he might say that he's not at all sure that Hayden isn't a genius.
> In your life have you met many John Nashes?
> I knew another J. F. Nash, but he's no longer with us. He was my father. I know another J. Nash, but not J. F. Nash. He's J. C. Nash. He's my son. I can't think of any others. Of course, J. Nash isn't a common name like J. Smith is.

I think that in these answers, which seem a little evasive (Nash never answers a question directly), we can see the myth he has always believed in: *to be unique*, and to be recognised as such in the eyes of the world. In the epilogue to the book *The Essential John Nash*, edited by Sylvia Nasar and Harold Kuhn, Nash wrote that the point of view of a person whose personal experience has become the subject of a book is different from that of the book's readers, and that in the overall experience of a person there is no "essential" and "non essential". According to Nash, the most beautiful thing is that a human being has the possibility to live and to experience, and that he can hope to be reincarnated or go to heaven when he has really come to the end and he is by now a part of history. It seems to me highly unusual that a person who defined a mathematical concept of rationality that is regularly used in the sciences should talk about reincarnation and heaven in the closing pages of a book that discusses his life and his Nobel Prize, and which contains the scientific works that made his famous the world over.

In closing, I believe that Nash in some sense rationally chose to get well. Of course, it may well have been that the passing years (even these bring some advantages) might have slowed down the chemical processes, as his long-time friend Harold Kuhn once defined them, that caused his illness in the first place, and it may be that the increasingly sophisticated and advanced treatments also did their part. However, the fact remains that there are many clues that suggest that there is a profound connection between his illness and the dreams, aspirations and objectives that he strived for. Not only, in spite of his having achieved great fame and received many honours late though they came, do we sense a kind of regret in Nash that he wasn't who he would have liked to have been, but we hear the kind of tragic desperation of a person who longs for immortal fame with all his might, and who, even though gifted with exceptional intellectual powers, has never lived up to his own expectations. To be sure, Nash's life was a particularly hard one, not only for

---

[1]Nash is right, it is always difficult to talk about one's self, especially in response to a question that is not overly acute like this one and the next.

himself but for those close to him as well, such as his wife, who, even though she had asked for a divorce, especially as a way to protect her child, never abandoned him, and today are married again. Nash is an exceptional person, but unfortunately in his own eyes, he was not and is not good enough.

Game Theory

In this book we talk about von Neumann and Nash, two giants of twentieth-century mathematics with two different kinds of temperament, two different stories, two completely different lives, but who nevertheless have many things in common. Not only, obviously, that for a certain time they were both at the Institute for Advanced Study in Princeton, one of America's temples to science, but also the fact that their contributions to mathematics, which developed in different areas, have one fundamental thing in common: both were considered fathers of a new theory, called Game Theory, a part of mathematics that is so new that it can't be taken for granted that even those who are interested in and up to date about the sciences know exactly what it is. Of course, we can reasonably guess that it deals in an intelligent way with how to play games that are popular all over the world, from chess to poker, to name only two. It is no coincidence that the first famous result refers to chess, and that Nash's doctoral thesis, the work for which he was awarded the Nobel Prize, contains the study of a simplified model for poker. It would be, however, an obvious mistake to think that this can be used only for analysing games. In reality, the best way to understand what game theory is about is to ask ourselves why games are so important. The answer is simple: games are important because they are a symbolic but very effective way of describing situations that life presents us with on a daily basis. In a game there are various people (players or agents) who have to make choices in accordance with certain rules. The set of their joint choices generates an outcome of the game itself, and the agents prefer certain outcomes over others. If a game is described in this way, then you can see that our lives are like games, and that we make decisions to obtain results that depend not only on our behaviour but on others as well. Studying these mechanisms means trying to shed light on the significance of rational behaviour, and trying to understand economic mechanisms, political theory, and psychology. Not only this, but if we hypothesise that there is also a kind of rationality in interactive behaviour even on the part of beings who can't be defined as intelligent, at least from the point of view of human intelligence, then we naturally see that this theory has applications in other areas as well, such as biology, genetics and computer science. I said earlier that the initial hypothesis is that there are agents who have to make choices, and that the combination of the choices determines the possible outcomes of the game, some of which will be preferred by the players. We presume that the players are egotists, that is, that they will be exclusively interested in obtaining their own maximum benefit, and that they are rational, that is, that they can in some way understand the best way to behave. It is this last assumption, apparently innocent, which obscures the many problems of interpretation and which is probably the reason why game theory is so fascinating, because the problem of correctly defining rationality has always permeated human thought. Now that we have given this rapid introduction, in what follows I will briefly talk about the earliest developments in the theory, with emphasis above all on the contributions of von Neumann and Nash.

The first noteworthy result is attributed to Zermelo, who in 1913 published a theorem about chess, in which he essentially says, or they say he says,[2] that games that have a structure like that of chess are determinate.[3] This means that at least in situations that can be modelled, like chess, we have an unequivocal idea of what it means to be rational, since the result of the game is predictable a priori. Chess is a zero-sum game. In simple terms this means that the players have opposing interests, and this is by far the easiest situation to model. It is no coincidence that successive contributions always involved this kind of situation, but with more complicated hypotheses. For example, we can understand that a game like rock-scissors-paper is of a different nature than chess: in chess the moves are in succession, and both players know how the whole game is played out; in rock-scissors-paper, the moves are made at the same time, so that when a player makes a move he doesn't know what his opponent will do, and vice versa. And in effect, even with two perfectly rational players, it is difficult to say that the outcome of any individual game will always be the same. However, it is pretty easy to intuit, especially in the case of a game that is played repeatedly, that there are more or less effective and intelligent ways of playing: for example, if I consistently make the next higher move over that of the previous stage, then my opponent will soon come to understand this strategy and can use it to his advantage. So von Neumann introduced the concept of mixed strategies, and used the concept of expected utility to extend the idea of equilibrium to these games as well. Finally, he stated a fundamental result, the so-called minimax theorem, which states that every finite two-person zero-sum game admits equilibrium, in mixed strategies. Thus even a game without apparent equilibrium shouldn't be played randomly or by trying to understand the opponent's psychology (this was E. Borel's idea, who was in any case sceptical about the possibility of achieving a minimax theorem). As the theorem says, there exists an "optimal" way for both players to assign a probability to each pure strategy (that is, a mixed strategy). Not only that, but since the players are rational, they both know the situation and thus such a game turns out to be determinate: its outcome is always and inevitably the same. Naturally this should be understood in a probabilistic way: if I play rock-scissors-paper against another player a sufficiently high number of times, then I know that at the end we will be substantially tied, and the only intelligent way to play is to trust in random choice, that is, random as understood in terms of certain probabilities. In the case of rock-scissors-paper, for instance, this

---

[2]Curiously, very curiously, the theorem published by Zermelo, although it does deal with chess, absolutely does not say what the experts say it says, even the most expert of experts.

[3]It isn't easy to explain in a few lines what this means and what the implications are. One way to say it is that any game played by two perfectly rational players would always end up the same way, meaning that either the two colours always tie, or that the same colour always wins.

means that I take a die and if it comes up as one or two, I play paper, three or four I play scissors, and five or six and I play rock.[4] Games that are strictly competitive have two fundamental characteristics. The first is that the players can calculate their equilibrium strategies independently: they must certainly bear in mind that the other player exists, but they don't need to know what he will do in order to achieve their best result. As a consequence, the second fundamental characteristic is that when there are various equilibrium configurations, even if they have two equilibrium points in mind, the results of their actions is an equilibrium, different from both the previous ones, but still assigning the same quantity to both. To explain this better, consider the following game, where on the contrary coordination between the players is essential to reach an equilibrium: if two friends want to spend an evening together but one wants to go to the opera and the other to the cinema, then it is obvious that while going together to the opera or going together to the cinema are both equilibria, if the two friends act independently without agreeing, they run the risk of finding themselves alone. Von Neumann's theorem is thus a great step forward in the theory of interactive decisions, perhaps the first systematic result in this sense. Clearly, the strictly competitive situation does not cover all the possible situations; on the contrary, we are much more likely to run across situations where the players can both benefit or both lose by making certain combined moves.[5] For this reason von Neumann thought about extending the theory that led to the publication of the book – co-authored with the Austrian economist Oskar Morgenstern – entitled Theory of Games and Economic Behaviour, which many consider to be the work that marked the birth of game theory, and which, along with the systematic presentation of the results of strictly competitive games, takes on the study of games that are not (necessarily) zero-sum using a totally new approach. The idea that is developed begins with the observation that a game can often be described by means of the behaviour of the coalitions that can be formed between players. Many examples illustrate this: a labour union, a political party, an association formed with the aim of obtaining advantages for its members. Once the game has been formalised by defining it as a function that associates to every possible coalition a set of utilities that the players of the coalition itself can obtain by coalescing, then a concept of a solution is formulated, that is, of a distribution of utilities among the individual players. In this case we speak of a cooperative approach to the theory, although we should clarify right away that the cooperation does not occur because of any decrease in the players' egotism. The hypothesis is simply that agreements which are somehow binding are stipulated because they are in everyone's best interests. The solution concept proposed by von Neumann and

---

[4]According to Einstein, God does not play dice. According to those working in game theory, God plays dice, but he plays intelligently.

[5]If this is not completely clear in terms of a board game, at least one with two players, it should be evident in the many examples we can find in economics, starting with the example of the only two coffeehouses in town which have to decide on the cost of a cup of coffee: it is in the best interests of each to keep the cost high, but if one lowers the price he earns more because he steals the customers of the other.

Morgenstern for this model of game is quite involved: there are many instances in which the solution is a set (not reducible to a single utility distribution vector) and that there are several solutions (making it difficult to interpret the concept itself a posteriori), nor were the authors able to establish whether or not there exist games without any solutions (the affirmative response would arrive years later). We are in the early 1950s now, when Nash bursts onto the scene as a student at Princeton's Institute for Advanced Study, in search of a good topic for his doctoral thesis and certainly fascinated by von Neumann's charisma. Nash doesn't hesitate to propose an alternative model to von Neumann's, which he himself defines as non-cooperative, in which the primitive data are the spaces for the players' strategies, each of which is assigned a utility function, which depend on all the players' choices. At the same time, Nash proposes a new solution concept, either of equilibrium, or of the idea of rationality, which today is called Nash equilibrium: a multistrategy – that is, a combination of strategies among the various players – is in equilibrium if no player, informed that the others intend to go along with the proposed multistrategy, has any interest in deviating from his own proposed strategy. In other words, if a referee says, "Claire, you choose strategy A and Camille, you choose strategy B", assuming that Claire effectively plays strategy A, then Camille has nothing to gain from changing strategy B (and vice versa, of course). On the other hand, Camilla can consistently assume that Claire will not deviate from strategy A, since she does not have interest in not following the recommendation, and conversely. In the same paper, Nash then proved an existence theorem. On the other hand, Nash says clearly in his thesis that his idea of equilibrium is not entirely new, since it was used much earlier and in a special case by Cournot. Be that as it may, the exceptional nature of Nash's contribution consists in his having formulated a model and formalised the equilibrium concept within it. After having published his thesis and the results contained in it, and after having developed almost at the same time a bargaining model between two agents, Nash essentially lost interest in game theory. It should also be mentioned that von Neumann did not accept Nash's model: his comment, not very generous by any account, is that it is only a new fixed point theorem. The later development of the theory would fully justify Nash: there is no doubt that the non-cooperative approach is today overwhelmingly prevalent, above all in the classic sectors where game theory is applied, economics first among them. From that time to the present, game theory has made giant strides forward. Not only that, but it has become one of the elite sciences, a fact demonstrated by the Nobel Prizes (1994, 2005 and 2008) awarded to those working in the field. Even more important, there is a more mature awareness of its values and its limits. It may be that there is no longer space for the enthusiasm with which the first results were met, or for the excessive optimism about what could be achieved by the systematic application of this discipline, but that is not necessarily bad. Game theory, like any good scientific theory, explains something, and often opens roads for research that are deeper and more interesting than the problems it actually solves. Today other theories which perhaps wouldn't have even been born had there been no game theory, have become very important, and the debates about their relative plusses and minuses are quite useful. One example is so-called behavioural economics. It is evident that the initial hypothesis – that of the players' perfect rationality – is a significant abstraction: for some this might pose a serious limit on the applicability of the theory as a whole. In reality, it isn't like that. Today the philosophers of science are inclined to think that the most valuable contribution of game theory is that of constituting a reference for determining how much the behaviour of agents can deviate from an optimal behaviour.

Without this reference it would be difficult to perform in-depth analysis that goes beyond the confines of purely qualitative considerations. I will close by noting that the paradigm of rationality defined by Nash's equilibrium concept raises more than a few questions, and a number of logical and philosophical dilemmas. I see this as a good sign. The most fertile ideas are not the ones that solve certain problems, even complicated ones, but those that raise new questions and open new horizons.

Ennio De Giorgi

Intuition and Rigour

Gianni Dal Maso

Ennio De Giorgi was born in Lecce on 8 February 1928. His father, Nicola, taught literature at the high school in Lecce, and was an expert in Arabic, history and geography; his mother, Stefania Scopinich, came from a family of seafarers from Lošinj in Croatia. His father died prematurely in 1930, but his mother, to whom Ennio was especially close, lived until 1988.

After graduating from the classical high school, in 1946 Ennio enrolled in engineering at the University of Rome. The next year, he switched to mathematics, receiving his degree in 1950 with Mauro Picone as his thesis advisor. Immediately afterwards, he was given a research grant to work at Picone's Istituto per le Applicazione del Calcolo (Institute for Applications of Calculation), and in 1951 he became Picone's assistant at the Mathematics Institute at the University of Rome.

Perimeter Theory and Hilbert's 19th Problem

In the years 1953–1955, De Giorgi obtained his first significant mathematical results in the theory of perimeters, a notion of a $(n-1)$-dimensional measure for oriented boundaries of n-dimensional sets introduced by Renato Caccioppoli. These results led to the proof of a isoperimetric inequality, published by De Giorgi in 1958: of all the sets of given perimeter, the hypersphere has the maximum n-dimensional volume.

In 1955, De Giorgi published a counterexample that showed the nonuniqueness of regular solutions of Cauchy's problem for linear partial differential equations with regular coefficients, a problem that had been unsolved for more than half a century. This brief article, lacking any bibliographic references, echoed throughout the mathematical world, arousing in particular the interest of Torsten Carleman and the admiration of Jean Leray. In 1966 Leray would construct some further "*contre-examples du type De Giorgi*".

C. Bartocci et al. (eds.), *Mathematical Lives,*
DOI 10.1007/978-3-642-13606-1_21, © Springer-Verlag Berlin Heidelberg 2011

Ennio De Giorgi

The most important result obtained by De Giorgi was the proof of Hölder continuity of the solutions of elliptic equations with measurable and limited coefficients (even when there is discontinuity in the coefficients). This result, obtained in 1955 and published in complete form in 1957, was the final – and perhaps most difficult – step towards solving the 19th problem posed by Hilbert in 1900: whether the solutions of regular minimum problems in calculus of variations for multiple integrals are regular, and even analytic in the case of analytic data.

The events leading up to the regularity theorem, recounted by Enrico Magenes during the commemoration of De Giorgi at the Accademia dei Lincei, unfolded in a blaze. In August 1955, while hiking near the Pordoi Pass, Guido Stampacchia told De Giorgi about the partial solution to Hilbert's 19th problem. De Giorgi must have immediately seen that it was possible to apply the results of his research on perimeters, particularly the isoperimetric property of the hypersphere, because in less than 2 months he was ready to present his proof of the theorem of regularity based on these tools at the Congress of the *Unione Matematica Italiana*.

This story sheds light on one aspect of the scientific personality of De Giorgi: lightening fast intuition joined by an exceptional ability to follow up with a proof formulated carefully down to the smallest details. The other aspect of his personality made evident by this story is his capacity to work on very difficult problems in almost total isolation.

The results regarding Hölder regularity had a significant influence on the theory of non-linear elliptic equations. The identical result for parabolic equations was proven in those same years by John F. Nash, Jr., using methods that were entirely different.

Some years later, in 1968, De Giorgi returned to the subject once again, using a counterexample to show that the same result does not apply to uniformly elliptic systems with discontinuous coefficients. Thus, if the question raised by Hilbert in the 19th problem is extended to vector functions, the answer is negative.

Minimal Surfaces

In 1958 De Giorgi became professor of mathematical analysis at the University of Messina. The next year, on the recommendation of Alessandro Faedo, he was called to the *Scuola Normale Superiore* in Pisa, where he held the chair in mathematical, algebraic and infinitesimal analysis for more than 40 years. Each year De Giorgi taught two courses, which usually met on Tuesdays and Thursdays from 11:00 until 1:00. The tone of the classes was very relaxed, with frequent interruptions for questions from those attending. Sometimes there was a 20 min break, and the whole class adjourned to a nearby café. Although somewhat fuzzy in the details, the classes were fascinating.

In 1960 the *Unione Matematica Italiana* awarded De Giorgi the newly established Premio Caccioppoli. During the 1960s his scientific work mainly regarded the theory of minimal surfaces. His main result was the proof of the analyticity almost everywhere of minimal boundaries in Euclidean spaces of an arbitrary dimension. This is a remarkable example of the great possibilities offered by the theory of perimeters in the calculus of variations. De Giorgi considered the result regarding the regularity of the minimal boundary to be a victory achieved in one of his most daunting scientific challenges.

The technique used in his proofs was immediately adopted by William K. Allard and Frederick J. Almgren to study the partial regularity of more general geometric objects and is today commonly used in contexts that are quite remote from those where it originated: non-linear elliptic and parabolic systems and equations, harmonic maps, problems of geometric evolution, etc.

In 1965 De Giorgi obtained an extension of Bernstein's theorem to dimension 3: all solutions of the equation for minimal surfaces defined for the entire three-dimensional Euclidean space are necessarily affine. This result was immediately extended up to dimension 7 by James Simons, who also constructed a locally minimal cone in dimension 8. De Giorgi then proved, in 1969, along with Enrico Bombieri and Enrico Giusti, that Simons' cone is also globally minimal. Further, using this cone, they constructed a non-affine solution to the equation of minimal surfaces defined on the entire Euclidean space of dimension 8. This surprising result showed that Bernstein's theorem could not be extended to spaces of a dimension higher than 7. In the same year, De Giorgi, along with Enrico Bombieri and Mario Miranda, proved the analyticity of the solutions to the equations of minimal surfaces in all spatial dimensions.

Between 1966 and 1973 De Giorgi enthusiastically accepted Giovanni Prodi's invitation to spend a month a year teaching at a the small university in Asmara run by Italian nuns.

In 1971, together with Lamberto Cattabriga, De Giorgi proved that, in dimension 2, every partial differential equation with constant coefficients and whose known term is real analytical has a real analytical solution, while in higher dimensions there are examples of even quite simple equations, such as that for heat, for which this property does not hold.

In 1973, the Accademia dei Lincei awarded De Giorgi the Prize of the President of the Republic.

G-Convergence

During the period from 1973 to 1985 De Giorgi developed the theory of G-convergence, conceived in order to provide a unified answer to the following question, which is present in many problems, both theoretical and applied: given a sequence F_k of functionals, defined in a suitable function space, does there exist a functional F such that the solutions to problems of minima for F_k converge towards the solutions of corresponding problems for F?

The starting point was the notion of the G-convergence of elliptic operators, introduced by Sergio Spagnolo in 1967–1968 and originally defined in terms of the convergence of solutions or corresponding equations. In 1973 De Giorgi and Spagnolo, reconsidered this notion from a variational point of view, making evident its connection to the convergence of the functionals of energy.

In an important paper published in 1975, De Giorgi went from the "operational" notion of G-convergence to one that was purely "variational". Instead of a sequence of differential equations, he considered a sequence of minimum problems for functionals in the calculus of variations. Without writing the corresponding Euler operators, De Giorgi established what could be considered as the variational limit of this sequence of problems, and at the same time obtained a result of compactness. This was the beginning of G-convergence.

The formal definition of this idea, together with the proof of its main properties, appeared a few months later in a paper co-authored with Tullio Franzoni. In the 10 years that followed, De Giorgi dedicated himself to developing the techniques of G-convergence and promoting its use in various asymptotic problems of the calculus of variation, such as the problems of homogenisation, of the reduction of dimension, of transition of phases, etc. De Giorgi himself, usually quite restrained when talking about his achievements, was very proud of this one, considering it to be a conceptual tool of great importance.

One characteristic of his work in this period was that it was the driving force behind a lively research group, introducing fruitful ideas and original techniques, and often leaving to others the task of developing them independently in various specific problems.

In 1983 De Giorgi gave a plenary lecture at the *International Congress of Mathematicians* in Warsaw. This was the era of Jaruzelski and *Solidarność*, and the congress, which had already been postponed by a year, was oppressed by a sense of pessimism. De Giorgi began his talk on G-convergence by expressing great admiration for Poland. On this same occasion he publicly expressed one of his most deeply held convictions, saying that in his opinion, man's thirst for knowledge was "a sign of the secret desire to see some ray of God's glory".

Equations of Evolution and Problems of Free Discontinuity

At the beginning of the 1980s, in a series of works with Antonio Marino and Mario Tosques, De Giorgi proposed a new method based on the idea of gradients for the study of equations of evolution of the gradient flow type. This method was applied to many problems of evolution with constraints that were non-convex and non-differentiable.

In 1983, during a solemn ceremony at the *Sorbonne*, De Giorgi was awarded an honorary degree in mathematics from the University of Paris.

In 1987 De Giorgi proposed, in a paper written with Luigi Ambrosio, a very general theory for the study of a new class of variational problems characterised by minimising volume and surface energies. In a later work he called this class "problems with free discontinuity", alluding to the fact that the set where the surface energies are concentrated is not fixed a priori and can often be represented by means of the set of points of discontinuity of a suitable auxiliary function. Surprisingly, in those same years David Mumford and Jayant Shah proposed, in the context of a variational approach to image recognition, a problem which was perfectly suited to De Giorgi's theory. The existence of solutions to this problem was proven by De Giorgi in 1989, in collaboration with Michele Carriero and Antonio Leaci.

Beginning at the end of the 1980s, De Giorgi was occupied with various problems of geometric evolution such as those of evolution by mean curvature, which require that the velocity normal to a surface must be proportional to its mean curvature at each point, and proposed various methods for defining weak solutions to the problem and calculating approximate solutions; his ideas would later be developed by various mathematicians.

In 1990 De Giorgi was awarded the prestigious Wolf Prize in Tel Aviv.

Foundations of Mathematics

Starting in the mid-1970s De Giorgi devoted his Wednesday classes to the foundations of mathematics, while his other classes continued to be dedicated to the calculus of variations or the geometric theory of measure. His approach to the foundations, which was non-reductionist in nature, required the identifying and analysing some concepts that were to be taken as foundations, without however forgetting that the infinite variety of the real can never be completely grasped, in much the same way as Hamlet warned Horatio, "There are more things between heaven and earth than are dreamt of in your philosophy", a reflection that De Giorgio adopted as a philosophy of his own. For his work on the foundations, the University of Lecce gave him, in 1992, an honorary degree in philosophy, of which he was particularly proud.

Social Commitment and Relations with Other Italian Mathematicians

Among De Giorgi's various social commitments, the one he felt most deeply about was undoubtedly that of human rights. This commitment, which he would work for until the very last days of his life, began in 1973 with the campaign to free Ukraine dissident Leonid Pliushch, unfairly imprisoned in a state mental hospital in Dniepropetrovsk. Thanks to the efforts of many scientists the world over, including Lipman Bers, Laurent Schwartz and De Giorgi, Pliushch became a symbol of the struggle for freedom of opinion. Finally, in 1976, Pliushch was freed. De Giorgi was able to involve hundreds of people with differing political opinions in this fight. Later he carried on with his work in defence of quite a number of people who were persecuted for political or religious reasons, becoming an active member of Amnesty International and never missing an opportunity to talk about and distribute the Universal Declaration of Human Rights.

In Italy De Giorgi had friends and students just about everywhere. He often travelled for seminars and conferences, especially to Pavia, Perugia, Naples, Trento, obviously in addition to Rome and Lecce. He was an assiduous participant in the congresses for the calculus of variations on Elba and at Villa Madruzzo in Trento, where he felt especially at home. He seemed tireless on these occasions, encouraging endless scientific discussions and throwing out new ideas or conjectures.

In spite of his being surrounded by colleagues, friends and students who admired him deeply, he remained quite modest. His office door was always open to anyone who wanted to discuss a mathematical problem with him. When this happened, he could sometimes appear a distracted listener, but he was always able to grasp the heart of the question and suggest new ways to address it, suggestions that always turned out to be effective.

He was a member of the most important scientific institutions, in particular the *Accademia dei Lincei* and the *Accademia Pontificia*, in which he was active until the end of his life. In 1995 he was called to be a member of the *Académie des Sciences* in Paris and the *National Academy of Sciences* in the United States.

De Giorgi was deeply religious. His attitude of being on an ongoing quest, his natural curiosity, his open-mindedness with respect to all ideas, even those farthest away from his own, led him to speak easily and constructively with others about religious topics.

Starting in 1988, when his first health problems appeared, De Giorgi spent long periods of time in Lecce, especially during the summers, with his sister Rosa, his brother Mario and their children and grandchildren, rediscovering the joys of family life. In September 1996 he was admitted to the hospital in Pisa where, in spite of several operations, he died on 25 October.

A Personal Recollection

I began to work on my degree thesis under the direction of Ennio De Giorgi in 1976. I had the extraordinary privilege of taking my first steps in scientific research under the guidance of the *Maestro* at exactly the time when he was developing his ideas on *G*-convergence. Thus I found myself in the enviable position of one who, without having done anything to deserve it and still wet behind the ears, is able to observe first hand an undertaking as exciting as the development of a new branch of the calculus of variations.

This is even truer in light of the fact that De Giorgi's habit was to let his students, even the youngest, in on the short and long term objectives of his research, going into them with us in great depth, patiently explaining when we had a hard time following him the reasons behind a conjecture or what techniques he thought were most suitable for proving it.

De Giorgi's style of working gave pride of place to opportunities for discussion, and not only as a method for judging the reasoning behind a problem, or for weighing the validity of a conjecture. In his conception of the work of a mathematician, informal discussion between friends of results achieved constituted an important part of scientific activity.

For De Giorgi, writing was something lifeless, though necessary of course, for giving results a definitive form and making them accessible through publication, but not as effective for spreading them as informal discussion between those who shared a lively interest in a given scientific problem.

Working in close contact with De Giorgi, I was able to experience directly the highest meaning of the expression "scientific school": a community of researchers bound together by common scientific interests, ready to discuss among themselves the results they had obtained and to exchange ideas as to which techniques to use to address the open questions; a community in which the knowledge and experience of the eldest members were transmitted to the youngest not only by means of formal occasions of classes and seminars, but above all through informal discussions and collaborative work.

At the time I was a student, and then later while perfecting my studies in Pisa, De Giorgi's school comprised, in addition of us students, various instructors at the University of Pisa, and a large number of collaborators from other universities who carried out their research in close contact with De Giorgi and visited frequently to discuss their results with him.

I recall that, for those of us who made up what you might call the Pisan nucleus of his school, the fundamental appointment was the Tuesday class, during which each year he taught a different subject, always dealing with stimulating unsolved problems; in those years he often dealt with G-convergence. These were classes of a very particular kind, in which the explanation of known results was the least important part, and always a function of acquiring the techniques to solve the open questions. These questions constituted the fundamental nucleus of the course. The most stimulating parts were De Giorgi's conjectures, usually quite detailed, as to the steps required to solve the open problems.

You might say that, while we attended the other classes to obtain information about the most important results of the past, we attended De Giorgi's class to obtain suggestions regarding the future. The problems raised during his classes were often taken up again during the discussions he had with small groups of people in his office. There he was able, in a less formal way, to explain the details of his conjectures, and to clarify, broadly speaking, the ways of reasoning about them that he thought were plausible. I remember that he never once excluded the possibility that a conjecture of his might be false, limiting himself to observing that one of the most important aspects of the mathematician's work lay precisely in identifying the propositions that were meaningful, and whose truth, or possibly falsehood, was of significant consequence for a certain theory.

In reality the majority of De Giorgi's conjectures which grew out of definitive results proved to be true. These conjectures were the most invaluable suggestions that De Giorgi used to give to all those in his school. Very often he preferred not to occupy himself personally with the proofs, leaving this task to others, both because materially he did not have enough time for all the proofs of his many ideas, which were sometimes quite complicated, and because he believed that it was much more important for him to indicate the directions that scientific research should follow.

De Giorgi was always open and available to listen to other mathematicians, both students and the older members of his school who came to tell him about their results and projects, as well as the many visitors who wanted to speak with him, from the most famous mathematicians in the world to those who had merely won a scholarship and wanted to ask his advice. I never remember him refusing to speak to anyone.

In reality these discussions with other mathematicians were a way to keeping himself up to date with the progress of others, without having to spend hours and hours in the library reading mathematics journals. Frequently only a mention of a new result that he had never read the proof of was enough to allow him to rapidly reconstruct it in his own personal way, sometimes shedding light on aspects of the research that had gone unnoticed by the author himself.

Quite often it appeared that he only listened distractedly to the explanation of a result of someone else, but in reality the significant details never escaped him. The interesting part of a discussion with him usually began only after you had finished telling him about the results obtained up to that time. At that point, even if he had seemed distracted at first, De Giorgi would suddenly become animated and began right away to provide interesting suggestions about possible new results that could be further deduced from those just obtained. If you didn't know his habits, you might think that he but little appreciated the work done. In fact, this wasn't the case, and De Giorgi's attitude towards his own work was just the same: once he had obtained a result, it continued to be interesting to him only in so far as it could be used as a point of departure for studying new problems that were still unsolved.

De Giorgi had an enormous influence on the training of several generations of students, who learned from his teaching and his example a special way of "doing mathematics", one which takes its cues from meaningful model problems, often (but not always) suggested by questions of applications, and in any case from stimulating mathematical difficulties, and resolves them by situating them within a broader theoretical context without, however, ever losing sight of the concrete problem, and thus makes it possible to explain its mathematical aspects in a satisfying way.

Laurent Schwartz

Political Commitment and Mathematical Rigour

Angelo Guerraggio

Mathematics, politics and butterflies were the three great loves of Laurent Schwartz, "father" of distributions, political militant unfailingly committed to the elimination of all oppression, and extraordinary butterfly "hunter" (with a collection of over 20,000 specimens).

Mathematics is passion and rigour; politics is justice; a world without butterflies would be very sad indeed.

From a strictly mathematical point of view, Laurent Schwartz's reputation is tied to the theory of distributions. He is also known, however, as an intellectual who lived through many of the great events of the second half of the twentieth century. All eras have a beginning and an end. Schwartz is one of the most luminous symbols of the era – very European, and very French – of the committed intellectual. Jean-Paul Sartre, Simone de Beauvoir, Luc Montagnier, Pierre Vidal-Naquet, François Mauriac, Yves Montand, Simone Signoret and others were all figures whose work was meaningful within the context of arenas that were much broader than their own particular field. What was important was politics, belonging (to the progressive front) and the commitment to "*à lutter pour les opprimeés, pour les droits de l'homme et les droits des peuples*" ("struggling for the oppressed, for human rights, and the rights of peoples").

It is not by chance that Schwartz's autobiography – from which all the quotations in this article are taken – is entitled *Un mathématicien aux prises avec le siècle* (English translation, *A Mathematician Grappling with his Century*, Birkhäuser, 2001). He begins by saying: "I am a mathematician. Mathematics has filled my life. . . . I have thought about the role of mathematics, research and teaching, in my life and the lives of others; I have pondered on the mental processes of research". He goes on to say, "In journalistic circles, which are always about two centuries late, it is still the custom to use the word "intellectuals" only for people in literature and the arts. When they talk about "the intellectuals" they refer only to them. It is true that literature and social sciences are essential to today's society, but scientific intelligence also has a fundamental function. It is not merely a series of automatic operations, but a grand discipline of the spirit, a culture and a form of thought which constantly transforms knowledge and society". And he concludes, "Mathematicians

C. Bartocci et al. (eds.), *Mathematical Lives*,
DOI 10.1007/978-3-642-13606-1_22, © Springer-Verlag Berlin Heidelberg 2011

transport their rigorous reasoning into situations of daily life. Mathematical discovery is subversive and always ready to overthrow taboos, and it depends very little on established power".

Laurent Schwartz

Schwartz was trained at the *École Normale Supérieure* – where almost all French winners of the Fields Medal were trained, with the notable exception of A. Grothendieck – and he had the opportunity to attend lectures and seminars by mathematicians of the calibre of Fréchet, Montel, Borel, Denjoy, Julia, Élie Cartan, and others. But the years between the two World Wars were a special time. The young students were aware that – if they succeeded in passing 2 or 3 years of *classes préparatoires* and an extremely selective entrance examination – they were entering a *grand école* (instituted by the *French National Convention* in 1974) and a national elite destined for advanced research and management. However, many were dissatisfied with the way subjects were taught, especially mathematics. French mathematics came out of the First World War in tatters. Now, in the 1930s, the old Masters were even older. There was no intermediate level of "eggheads" who were "full of fight", open to new ideas and capable of speaking directly to young people. Schwartz repeatedly complains about a teaching that was fragmentary and lacking in great ideas: "Mathematics appeared to be a finished subject. ... It seemed there wasn't much left to do. And we didn't even have an idea of that little bit. ... I clearly perceived something lacking in everything we learned; the absence of some unifying thread". His criticism extended to the textbooks and monographs used by the *normaliens*. The *romans* – as the new generation called the volumes of the *Collection Borel* – on the one hand said too much and went on at length about details that were of trifling interest, while on the other hand they lacked the necessary rigour and shed no light on the important ideas.

For Schwartz, the encounter with Bourbaki was determinant: "...what was lacking for me was Bourbaki. I needed their inspiration to really become a mathematician". He became aware that there is no fundamental difference between the abstract and the concrete: a concrete object is only an abstract object that we have become accustomed to. He found the right space for his need for rigour, finally arriving at the conclusion – following on the heels of André Weil – that the Italian school of algebraic geometry was one of the most significant examples of the lack of rigour of the previous generations.

The encounter with the Bourbaki group took place in 1940. Schwartz had finished his studies at the *École Normale Supérieure* in 1937; he had married Marie-Hélène Lévy, daughter of Paul Lévy, one of the founding fathers of modern calculus of probability; he had started a long period of service in the military, made longer by the eruption of the Second World War. Schwartz's generation was one which, right in the best days of their lives, was forced to come to terms with tragic events all concentrated in a few short months: the annexation of Austria, the Munich agreement, the occupation of Czechoslovakia, the agreement between Hitler and the Soviets, the Spanish civil war and the victory of Franco, the Nazi invasion of Poland, the outbreak of World War II, the Vichy government, and more.

It was in fact years after his apprenticeship at the *École Normale Supérieure*, only after the liberation of France in October 1944, that Schwartz was able to turn his attentions once again to mathematics. Dieudonné and Delsarte called him to Nancy, where his students would include J. L. Lions, B. Malgrange and A. Grothendieck. After that, Choquet and Denjoy convinced him to move to Paris, where he had the chance to create an even stronger "school". Schwartz taught at the university, and then later, starting in 1969, at the *École Polytechnique*.

The distributions date from 1944, with the name chosen for their physical significance, in that they can be interpreted as distributions of electrical or magnetic charges. Schwartz speaks about the night of discovery as "a marvellous night, the most beautiful night of my life":

> In my youth I used to have insomnias lasting several hours and never took sleeping pills. I remained in my bed, the light off and without writing, did mathematics. My inventive energy was redoubled and I advanced rapidly without tiring. I felt entirely free, without any of the brakes imposed by my daily life and writing. After some hours ... especially if an unexpected difficulty came up ... I would stop and sleep until morning. I would be tired but happy for the whole of the following day. ... On this particular night I felt sure of myself and filled with a sense of exaltation. I lost no time in rushing to explain everything to Cartan who ... lived next door. He was enthusiastic: "There you are. You've just resolved all the difficulties of differentiation. Now we'll never again have functions without derivatives".

In essence, distributions constitute a generalisation of the concept of function, which in fact resolves the problem of derivation (extending its calculation and conserving its principle rules). A distribution is always derivable, sometimes an infinite number of times, and its derivatives also represent distributions. Precisely because there are no exceptions to the process of derivation, the theory of distribution forms the most natural context in which to situate any differential problem; once the solution is found, it is equally natural to ask whether the distribution found is, in particular, a function.

This procedure is not new in mathematics. For example, the search for the real solutions to an equation can be collocated in the field of complex numbers, checking later to see if the solutions found are real or not, that is, if the complex numbers were used to express real solutions.

The theory of distributions itself has numerous precursors, which Schwartz took pains to list, recalling the constant need to give meaning to the operation of derivation, or precise contributions in this direction – perhaps at first with insufficient rigour – or specific "technical" contributions: Heaviside's symbolic calculus, a 1912 paper by Peano, Dirac's function, Bochner's generalised solutions and Leray's weak solutions, Sobolev's functions, de Rham currents, etc. Schwartz's distributions was heir to all these attempts, which now appeared to be individual aspects of an organic and rigorous theory that Schwartz himself unhesitatingly claimed had "deeply changed the whole nature of analysis".

In recognition of Schwartz's contribution, in 1950 he was awarded the Fields Medal. But there was one, not insignificant problem. The Fields Medal was presented during the opening ceremonies of the International Congress of Mathematicians, which was to be held that year in the United States. Schwartz's political activity was well-known on that side of the Atlantic, perhaps even more so than his mathematics, not least because of the McCarthyism that pervaded America in those years. Schwartz ran a serious risk of not being issued a visa. This led the French mathematicians, whose spokesman was Henri Cartan, then president of the *Société mathématique de France*, to threaten to boycott the international congress, followed by many American mathematicians who were equally intransigent. The conflict was a hard one, in part because the French nominated as head of their delegation the elderly Jacques Hadamard, a relative of Schwartz, one of the most esteemed mathematicians, but also a communist sympathiser. In the end, not even the US State Department was enough; President Truman himself had to intervene (the Korean war was in full swing): Schwartz and Hadamard were issued visas, the international congress was salvaged, and Schwartz could personally receive the Fields Medal. (In 1972 he was elected to the *Académie des Sciences*.)

We have touched on the public and political dimension of Schwartz's life, which would soon become the dominant one (with the inevitable repercussions on his research). Schwartz would in any case always remain a mathematician, one with a great love for his discipline: "... mathematics are – not the queen of sciences as has been said too often; there is no queen of the sciences – but a very great, true and magnificent science". He feels the same way about research and teaching. Thus regarding research, he writes,

> Every time I try to read mathematics or listen to a talk I feel as though it is an assault. It's as though they were trying to destroy my castle. ... It almost seems as though my castle is an

obstacle to development. But I don't believe so . . . I feel a kind of imperialist desire for total knowledge, not only for mathematics but also for sciences and everything to do with life and society. For me everything ought to be perfectly logical. I can't tolerate fuzziness. If I don't know a theory well, I feel as though I don't know it at all. I have difficulty accepting half-measures.

Regarding teaching, he says, "when I've had the joy of teaching my students a beautiful theorem, I often prolong this pleasure by saying the lesson over to myself, maybe even out loud, on the way home. *Quelle sensualité*". Schwartz doesn't even hesitate to introduce the topics and styles most dear to the Bourbakis in the courses for engineers, agreeing to transfer to the École Polytechnique only after he had received ample assurance that he would be allowed to introduce radical reforms in the engineering program. Whatever the underlying cause, whether the novelty of "modern" teaching or the fascination of rigorous and strict logic, or perhaps the charm of a great instructor, the fact remains that the "new" way of teaching analysis was hugely popular among the students of the École Polytechnique.

The students' appreciation is even more impressive if we bear in mind that Schwartz was light-years away from being easy going or demagogic. He constantly ranted against the impoverishment of mathematical education and the destructive tendencies towards a sham egalitarianism: "Over the past 30 years there has been a real improvement in education thanks to the fact that schooling has become mandatory until 16, yet the renewal of the elite seems to me to be less effective than it used to be". We find him reiterating these ideas in many of his works, such as the article entitled "L'Enseignement et le développement scientifique" written in 1981, as part of an overall "snapshot" of the French social system sponsored by Prime Minister Pierre Mauroy or the 1984 book *Pour sauver l'Université*, or the by-laws of an association – QSF, "Qualité de la science française" – for the protection of the values of scientific culture and its transmission. But now it is time to speak in particular of Schwartz as "politician".

Schwartz's militancy began very early: "I had become a Trotskyist in 1936 and remained very militant until 1947". He took his militancy very seriously, both during the year of Germany occupancy – with all the risks inherent in that stand, including deportation – and in the years immediately after the war, vehemently objecting to the Stalinist orthodoxy of the French communist party. Schwartz was a "grass roots militant", distributing door-to-door *La Verite*, the newsletter of the "Internationalist Workers Party, French Section of the IV International", happiest when he succeeded in selling his 50 copies each Sunday morning. In the final 2 years, he was even elected to the party's Central Committee, and had good prospects of being elected National Secretary (providing, of course, that he give up his work as a mathematician): "The idea of becoming an apparatchik was not exalting, but even though this may seem surprising today, I did not radically reject it until after a couple of weeks of reflection. At that time I was a firmly convinced political militant".

Schwartz left the Trotskyite party because he saw it was becoming sclerotic. It was a party that was outside reality, one that made no progress in spreading its political creed, and was torn apart by internal bickering.

The experience, however, brief though it was, left lasting marks on both his personality and his future political choices; Schwartz was repeated mocked as an "old Trotskyite". In later years, he would join various political–cultural groups sponsored by Sartre, and finally the Socialist party. In any case he remained a man of the left (even the far and radical left), with a strongly ethical conception of politics, never losing sight of two particular points of reference: "There are two subjects on which my Trotskyite ideas have not changed at all: internationalism and anticolonialism".

The two points would become immediate evident when the serious crisis in Algeria broke out. It was 1954. Algeria, a French colony, began its struggle for independence, as did many other third world countries at the time. Schwartz had no doubts about what was right – among the French leftists, the position was a given – and he at once took a position in favour of the autonomy of the Algerian people.

His commitment became ardent when the Audin scandal erupted. Maurice Audin was a young Frenchman, a mathematics student, and communist, who Schwartz had met during the preparation of Audin's thesis – thanks to an introduction by René de Possel, one of the "elders" of the Bourbaki group. On 11 August 1957 Audin was arrested in Algiers by French parachute forces, and was tortured and beaten to death. For the whole month of June, even though Audin had died on 21 June, the French authorities continued to provide assurances that he was being held in a safe place; then, confronted with insistence by his wife and friends, they invented a tragic accident to justify his death. Schwartz was one of those who received a telegram from Audin's wife asking for help in bringing her husband's fate to light. Thus was born the Audin Committee, with Schwartz himself as president (the committee would later become a point of reference in the fight against torture in general). This was the start of a long and hard struggle to raise public awareness and put pressure on the French government and on General de Gaulle. With de Possel and the support of the academic authorities, Schwartz decided that Audin was to defend his thesis – in absentia – at the Faculty of Sciences in Paris, with de Possel undertaking the defence and a deliberation following. The impact on public opinion was remarkable. The weekly *L'Express* put Schwartz on the cover. The tension and protests reached their climax in 1960, at the time of the Manifesto of the 121: Schwartz was one of the signers of a manifesto that exhorted young people in France to an act of insubordination, to become conscientious objectors and refuse to support the war in Algeria. Again a shockwave echoed through France; there were no lack of consequences. Schwartz was suspended from the École Polytechnique and from the honour of teaching there, because the École had a long military tradition. His response and his indignation, even though many years have passed, are still moving, "If I signed the declaration of the 121 it was partly because for years I have seen torture go unpunished and torturers rewarded. My student Maurice Audin was tortured and assassinated in June 1957, and you, Mr. Minister, signed the exceptional promotion of Captain Charbonnier to the grade of officer in the Legion of Honor and the promotion of Captain Faulques to the grade of commander in the Legion of Honor. I repeat:

"Honor". Coming from a Minister who has accepted these responsibilities, considerations on the subject of "honor" cannot but leave me cold".

Schwartz's commitment to the struggle in Algeria would have other repercussions on his personal life as well in addition to the "simple" temporary removal from the École Polytechnique. In February 1962 his son Marc-André, not yet 20-years old at the time, was kidnapped by a group of commandos of "French Algeria" and spent 2 days in captivity. It was a terrible shock for the boy, made worse by rumours that started to spread immediately after his release that the "fake" kidnapping had been ideated and orchestrated by the victim himself. Marc-André never completely got over it, and finally, in 1971, after several failed attempts, shot himself in the temple.

To "disintoxicate" himself from the Algeria experience, in 1962 Schwartz accepted an invitation to spend a year in the United States, along with his family. But his civil and political commitment went with him. In the mid-1960s the Vietnam war broke out, and once again Schwartz did not hesitate to side against imperialism, this time American imperialism. It all began when he organised "Six Hours" of solidarity with the Vietnamese people, which was successful beyond all expectations. Then there was the Russell Tribunal, and Schwartz's acceptance of Bertrand Russell's invitation to be one of the twenty notables who comprised the jury of this tribunal against war crimes. Schwartz visited Vietnam several times, and even met with Ho Chi Minh, "My fight for the freedom of Vietnam was the longest fight of my existence. I love and will always love Vietnam, its landscapes, its extraordinary people, its bicycles. I am a little bit Vietnamese. Meeting a Vietnamese person or hearing Vietnamese spoken in the bus (even though I don't know the language) makes me inexplicably happy. My sentimental fiber vibrates for the country ... The Vietnamese do not forget me and many students write to me, calling me 'the godfather of all Vietnamese'".

This commitment to human rights and the rights of peoples was one to which Schwartz dedicated a large part of his life. But he still takes a certain ironic tone when he lists the infinite series of committees, associations and groups that invited him to be president. Algeria, Vietnam, Afghanistan and so on, up to the Committee of Mathematicians in defence of human rights.

The story of the Committee of Mathematicians began with the case of Leonid Pliushch, a Soviet dissident, mathematician, and translator into Russian of the first volume of Bourbaki's *Théorie des ensembles*. Pliushch was arrested by Soviet police in September 1972 and then forcefully committed to a psychiatric hospital. Schwartz called on the whole mathematics community to bring pressure to bear in support of Pliushch, with the active help of Henri Cartan, Claude Chevalley and Jean-Pierre Serra. Mathematicians the world over mobilised. Lucio Lombardo Radice, a member of the Central Committee of the communist party and Ennio De Giorgi were among those most active in Italy. Finally, in February 1976, Pliushch was freed.

Following Pliushch, there was Massera, a mathematician and communist from Uruguay who was imprisoned by his country's army. Joining in the fight to free Massera was Jean Dieudonné, a member of the right, who was vehemently anti-communist but who personally assisted in the negotiations to free his South American colleague, and who wrote an article on Massera's mathematical work. Still another case was Sakharov – a case in which René Thom intervened – and many other cases in which the Committee of Mathematicians intervened to defend mathematicians all over the world who were accused and arrested because their political opinions were illegal.

What more can we say? There remains only the butterflies ... Schwartz had learned to love and collect them in the "garden of Eden" that was his summer vacation home when he was a child. "Pesticides have caused the disappearance of butterflies in the countryside. Ecologists have never become interested in the disappearance of the butterflies. But a world without butterflies would be a sad place".

"Who Is" Laurent Schwartz

Laurent Schwartz (1915–2002) attended the École Normale Supérieure, where he finished his studies in 1937. The following year he married Marie-Hélène, daughter of Paul Lévy (one of the founders of the modern calculus of probabilities), with whom he had two children, a son Marc-André and a daughter Claudine. His son committed suicide in 1971, the result of the life-long trauma following his kidnapping at the hands of French nationalists seeking revenge on his father for his commitment to anti-colonialism and support of the Algerians seeking independence.

His academic life took Schwartz to Grenoble, Nancy and Paris. In Paris in 1969 he left the university to teach fulltime at the École Polytechnique, where he was personally involved in the reform of the engineering curriculum and where he remained until his retirement. The École Polytechnique honoured his memory with an important conference that took place in Palaiseau in July 2003.

His mathematical research was strongly influenced by his encounter with the Bourbaki group in 1940. The theory of distributions was born in Paris, in early November 1944. He was awarded the Fields Medal in 1950.

Noteworthy among Schwartz's many civic and political commitments was his devotion to scientific education and culture. In 1981 he was charged by Prime Minister Pierre Mauroy to oversee – as part of the project aimed at providing an overview of the situation in France – the fourth volume dedicated to scientific teaching and development. Schwartz fought relentlessly against egalitarianism and laxity in education, concepts which he defined as destructive. He always believed that the acknowledgment of the existence of differing levels of merit was a driving force that tended to push everything to a higher level, while egalitarianism tended to a lowering of levels.

René Thom

The Conflict and Genesis of Forms

Renato Betti

Born in 1923 in Monbéliard, France, René Thom, an interdisciplinary mathematician by vocation and perhaps by training, which took place within the great French school of mathematics, was also remarkably adept at developing the inherent technical aspects of the great specialisations. With the results achieved while still a young man he won the Fields Medal, the highest international recognition for mathematicians. By laying out the general concepts, he paved the way to an original attempt to apply mathematics to natural phenomena, today known as "catastrophe theory".

After earning his degree in mathematics at the Paris École Normale Supérieure in 1946, he became a researcher in Strasbourg and continued his studies with Henri Cartan, earning his Ph.D. in 1951 with Cartan as his thesis advisor. The thesis, entitled "Fibre spaces in spheres and Steenrod squares" concerned the topological invariance of certain classes of manifolds and formed the basis for the "theory of cobordism", that is, a general theory of forms and of their stable "singularities": here there already appear the first notions that, when later developed, led him to the Fields Medal, presented to him at the International Congress in Edinburgh in 1958.

Following the award of the prestigious prize was the move in 1963 to the Institut des Hautes Études Scientifiques in Bur-sur-Yvette, and the opportunity to leave the world of specialized mathematics to address more general notions, such as the theory of morphogenesis, a subject that would lead him to a very general form of "philosophical" biology and the search for "a common grammar of the most disparate phenomena", to use Thom's own expression.

"Catastrophe theory" was the fruit of his reflections. Going beyond the axiom of permanence of effects on continua, and arriving at the consequent acceptance of the fact that marginal causal variations can lead to tangible effects, led to the notion of "catastrophe": a sudden shift that occurs in a system – whether physical, biological, social, or even linguistic – that is also subject to regularly variable conditions. From here the search for non-equivalent canonical forms, and then to the theorem of classification that identifies seven elementary catastrophes.

The theory, along with its reasoning, models and general concepts, would be developed in two major works, *Structural Stability and Morphogenesis* (1972), and

C. Bartocci et al. (eds.), *Mathematical Lives*,
DOI 10.1007/978-3-642-13606-1_23, © Springer-Verlag Berlin Heidelberg 2011

Mathematical Models of Morphogenesis (1983), which led to further studies and applications by mathematicians as well as nonmathematicians.

René Thom

There was no lack of controversy, often fueled by Thom himself, who had admittedly left the Bourbaki group he had belonged to by training and by culture, proclaiming, above all at the level of education, the importance of intuition and of mathematics "that can be seen and touched", but at the same time he was highly critical of those who wanted to move immediately to the applications of his theory – to numerical and quantitative predictions, from an interpretation that he considered hermeneutic and qualitative. For Thom, promising results from catastrophe theory also arrived from "metaphorical" considerations, when it is thought of as a rigorous theory of analogy in which the richness of language is no less important than the classification of forms.

The scientific aspects of Thom's career, in which catastrophe theory led him to examine more broadly the topics of the stability of organized systems and their classification, was closely connected with its more philosophical aspect, where he tended to broaden the vision of the world made possible by mathematics – in this case, and particularly in the case of differential topology, which Thom himself had helped to found. The morphogenesis of structures concerns their creation, growth and end. It is a subject that has famous precedents, beginning with Aristotle. Thom's work was considered a continuation of that of D'Arcy Thompson, who, at the turn of the twentieth century, had begun to apply principles of mathematics – especially geometry – to biological systems, thus offering to theory a very solid mathematical apparatus.

In the natural world of living beings, as well as in the artificial world of man-made systems, there are recurring structural forms that are the result of necessity, not chance. In Thom's "a priori metaphysics" there is a conviction that the world is not chaos, but rather and ordered *cosmos*. Its forms are distinct and separate from each other. What is attempted is finding in a rigorous way the general schemes that make it possible to explain the genesis, in the belief that the world is intrinsically rational, and with a view of occurrences that is substantially determinist.

J.L. Borges, The Dream (El sueño)

Night sets on us its magic task:
to ravel out the world, the endless branchings
of cause and effect, which lose themselves in time's
unfathomed vertigo. Night demands that every night
you forget your name, your blood, and those who bore you,
each human word, each tear, and everything
that being awake has ever taught you – geometry's
imaginary point, the line, the plane, the cube,
the pyramid, the cylinder, the sea and waves,
the coolness of clean sheets, gardens, empires,
the Caesars, Shakespeare, and, what's hardest of all,
the one you love. Strange to think that a pill
blotting the cosmos out, lets chaos in.

Jorge Luis Borges

(translated from the Spanish by Norman Thomas di Giovanni; http://www.
digiovanni.co.uk/index.php)

C. Bartocci et al. (eds.), *Mathematical Lives*,
DOI 10.1007/978-3-642-13606-1_24, © Springer-Verlag Berlin Heidelberg 2011

Alexander Grothendieck: Enthusiasm and Creativity

Luca Barbieri Viale

C'est à celui en toi qui sait être seul, à l'enfant, que je voudrai parler et à personne d'autre.[1]

Alexander Grothendieck was born in Berlin on 28 March 1928. His father, Sascha Shapiro, an anarchist originally from Russia, took an active part in the revolutionary movements first in Russia, and then in Germany, during the 1920s, where he met Hanka Grothendieck, Alexander's mother. After the Nazis came to power in Germany it was too dangerous for a Jewish revolutionary to stay there, and the couple moved to France, leaving Alexander in the care of a family near Hamburg. In 1936, during the Spanish Civil War, Sascha joined the anarchists in the resistance against Franco. In 1939 Alexander joined his parents in France, but Sascha was arrested and – partly as a consequence of the race laws enacted by the Vichy government in 1940 – sent to Auschwitz, where he died in 1942. Hanka and Alexander Grothendieck were also deported, but they escaped the holocaust. Alexander, separated from his mother, was able to attend high school at the Collège Cévenol in Chambon-sur-Lignon, lodging at the Secours Suisse, a hostel for refugee children, but he had to flee into the woods every time there was a Gestapo raid. He then enrolled at the University of Montpellier and in autumn 1948 he arrived in Paris with a letter of introduction to Élie Cartan. This led to his being accepted at the École Normale Supérieure as an *auditeur libre* for the 1948–1949 academic year, where he assisted in the debut of algebraic topology in the seminar taught by Henri Cartan (Élie's son). His earliest interests, however, were in functional analysis, and following Cartan's advice, he moved to Nancy. Under the guidance of J. Dieudonné and L. Schwartz, he earned his doctorate in 1953.

During his years in high school and university, Grothendieck never much enjoyed the courses and programs he attended, nor can it be said that he was a model student. His curiosity, coupled with a sense of dissatisfaction, drove him, not quite 20-years old, to develop on his own a theory of measurement and integration.

---

[1] It is to the one inside you who knows he is alone, to the child, that I wish to speak and to nobody else. *Récoltes et Semailles*, "Promenade, à travers une œuvre", p. 7.

C. Bartocci et al. (eds.), *Mathematical Lives*,
DOI 10.1007/978-3-642-13606-1_25, © Springer-Verlag Berlin Heidelberg 2011

When he arrived in Paris, he found it had already been written by Lebesgue. He said, "I learned then in solitude the thing that is essential in the art of mathematics – that which no master can really teach".[2] The "official" productive period of Grothendieck's life, as testified by an impressive mass of writings, is 1950 to 1970. While the research topics of the early 1950s were those of functional analysis, the great themes of algebraic geometry, its foundations, such as the redefinition of the concept of space itself, occupied the years 1957–1970.

In 1959, by now a professor at the newly created Institut des Hautes Études Scientifiques (IHES) in Bures, near Paris, Grothendieck taught a lively seminar in which he – in a magnificent display of generosity – shared and gave away his research ideas to students and colleagues, developing them with boundless enthusiasm and creativity. In these early years his frequent and intense contacts with Jean-Pierre Serre, traces of which are left to us in their correspondence, were a source of inspiration and mutual exchange of ideas. In the decade between 1959 and 1969 Grothendieck's ideas were mainly spread, on the one hand, through publications such as *Éléments de Géometrie Algébrique* (*EGA*) – edited in collaboration with Dieudonné – and with the help of the participants in the *Séminaire de Géométrie Algébrique* (*SGA*), and on the other hand, through *Exposés* at the Bourbaki seminars. According to Grothendieck's original idea, the *Séminaire* was considered as a preliminary form of the *Éléments* and was destined to be incorporated into it. The *Éléments* were initially published by the IHES in various weighty tomes. In 1966, Grothendieck was awarded the Fields Medal (the highest recognition a mathematician can receive).

In 1970 Grothendieck, then 42-years old, officially abandoned the scene. There were many reasons that led him to withdraw from the academic world, but certainly his radical anti-war stance was one that he declared openly. It had come to his attention that the IHES received funding from the defence ministry – and had received it for more than 30 years without his being aware of it – and his response was to desert the *Institut*, also taking away with him the publication of the *EGA* and the *SGA*, signing a contract for the new edition with Springer-Verlag. Knowing what it is like to live as a refugee, with a United Nations passport – his own original documents disappeared during the Nazi holocaust – he gave life to the pacifist and environmental movement named *Survivre*. Seen against the background of the major issues of those years, the Vietnam war and the proliferation of nuclear weapons – war and the stockpiling of weapons of mass destruction are still issues still quite pertinent today – Grothendieck's pacifism shows a significant shouldering of responsibility, not the kind that the institutions involved could ignore (even though these still today receive the same kind of funding). Following this decision, Grothendieck spent a couple of years at the Collège de France and then in Orsay before finally returning to the University of Montpellier in 1973. He refused the Crafoord Prize in 1988, the year of his retirement. In these last years, retiring to private life in the country near Mormoiron, having given up travelling, he dedicated

---

[2]*Récoltes et Semailles*, p. 5.

himself to correspondence and to the *Récoltes et Semailles*, a long diary-like narrative about his past as a mathematician, or as he says, a long meditation on "the internal adventure that was and is my life".[3]

Alexander Grothendieck

I received portions of the *Récoltes et Semailles* in 1991, along with a letter from Grothendieck in which he told me that Aldo Andreotti was "a good friend and a truly valuable person: I came to appreciate his peculiar qualities much more now than he has passed away than I did during the 1950s and 1960s when he was still alive". I don't know which Italian mathematicians worked with Grothendieck in those years; the Italian schools were slow to assimilate his methods in algebraic geometry, even though these were partly rooted in the work of Italians such as Severi and Barsotti.

The *Présentation des Thèmes* of the *Récoltes et Semailles* provided the valuable information – along with the letter I have just mentioned – for the sketch of his life given up to now, and for an outline for an overview of his mathematical thinking, to which we will now turn.

Grothendieck's excellence, his mathematical genius, is quite evident in his innate tendency to bring to the fore themes that are obviously crucial but which no one before him had made evident or acknowledged. His productivity had deep roots and expressed itself by means of language that was ever new, emerging like a flowing river of new *notions–abstractions* and *statements–formulations*. Quite frequently statements that sprang perfectly formulated from his fervid and

---

[3]*Récoltes et Semailles*, p. 8.

implacable imagination turned out to be the foundation of an entire theory that Grothendieck himself outlined, developed and followed through with; in other cases they are only sketched out.

If by mathematical dexterity we mean man's capacity to solve problems, then this tendency of his not just to find solutions to mathematical problems but to *create* mathematics, makes Grothendieck an extremely special and extravagant mathematician. The layman who approaches Grothendieck's mathematical work has to get past the usual concept of a mathematician as a problem solver and try instead to see mathematics as an art and the mathematician as an artist. Of course, mathematics is a very special kind of art, one in which *inventions* borrow from the proofs, that is, imagination has to harmonise with reason. The mathematician' works are theories in a weave, a design, that always make it possible to grasp a oneness in multiplicity. As Grothendieck himself wrote, "it is in this act of going beyond, not in remaining closed within a mandatory circle that we ourselves create, it is above all else in this solitary act that creation is found".[4]

For Grothendieck, mathematical theories are also opportunities for reflection in a lateral sense, and meditative exercises, a kind of contemplation that accompanies us on our internal adventure. Mathematics is thus a *yoga* that diversifies and multiplies into different theories but whose foundations are firmly united. The differentiation of these old and new themes is also interwoven with the history of the ideas that inspired them. According to Grothendieck, there are traditionally three aspects of things that are the objects of mathematical reflections: number, or the arithmetic aspect; measure, or the metric (or analytic) aspect; and shape, or the geometric aspect. "In the most part of the cases studied in mathematics, these three aspects are either present simultaneously or in intimate interaction".[5]

Let's look at some of the topics that algebraic geometry involves from Grothendieck's point of view. His was a perspective that favoured shape and structure and thus the geometric and arithmetic aspects, in a unifying vision that gave birth to a new geometry: *arithmetic geometry.*

> We can state that number is aimed at grasping the structure of the disparate or discrete parts: the systems, sometimes finite, formed of elements or objects that are isolated, if you will, in relation to each other, without any principle of continuous passage from one to the other. Magnitude, on the other hand, is the quality *par excellence*, susceptible to continuous variation; for this reason it is aimed at grasping structure and continuous phenomena: motions, spaces, variations of all kinds, force fields, etc. Thus arithmetic appears (more or less) as the science of discrete structures, and analysis as the science of continuous structures.
>
> As far as geometry is concerned, we can state that after more than 2,000 years it exists as a form of science in the modern sense of that term, it straddles the two kinds of structure, discrete and continuous. On the other hand, for a long time there was no real "divorce" between the two different kinds of geometries, one discrete and the other continuous. Instead, there were two different points of view about the investigation of the same geometric figures: one placed an emphasis on the discrete properties ... the other on the continuous properties. . . .

---

[4]*Récoltes et Semailles*, p. 6.

[5]*Récoltes et Semailles*, p. 26.

At the end of the 1800s there was a divorce, with the birth and the development of what was sometimes known as abstract (algebraic) geometry. Roughly speaking, its aim was to introduce, for every prime number p, an (algebraic) geometry of characteristic p, based on the (continuous) model of the (algebraic) geometry inherited from earlier centuries, but in a context that appeared, however, to be irreducibly discontinuous, discrete. These new geometrical objects became increasingly important at the beginning of the 1900s, and this in particular, given the close connection with arithmetic ... would seem to be one of the guiding ideas in the work of André Weil.... It is in this spirit that he formulated, in 1949, his celebrated Weil conjectures. Conjectures that are absolutely astounding, if truth be told, making it possible for us to see, by means of these new discrete kinds of varieties (or spaces), the possibility of certain kinds of constructions and topics that up to that time had seems conceivable only in the context of those spaces that the analysts deemed worthy of being called by that name....

It is possible to believe that the new geometry is above all a synthesis of these two worlds ... the arithmetic world ... and the world of continuous magnitudes. In this new vision, the two worlds that were once separate, now form a single world.[6]

This unifying vision is embodied in the concepts of *scheme* and *topos*, revealing hidden structures: the geometrical richness of the discrete world is brought to light in all of its beauty and detail, thus making possible for Grothendieck himself and his student, Pierre Deligne, to prove the so-called Weil conjectures.

The concept of scheme constitutes an enlargement or generalisation of the concept of algebraic variety as it had been studied by the Italian and German schools in the early years of the 1900s. Grothendieck's idea of scheme and the basic ideas of a scheme theory, by means of the concept of maps, that is, by a suitable transformation (or morphism) of schemes, goes back to the years 1957–1958 and were briefly illustrated at the International Congress of Mathematicians in Edinburgh in 1958. It was precisely the concept of *sheaf* – already introduced and studied by Leray, Henri Cartan and Serre – that turned out to be essential because it made it possible to reconstruct a global datum starting from an open set of locally defined data, and thus making it possible to apply continuous reasoning in a discrete context.

While algebraic geometry is the study of polynomial equations and the geometric loci that they define, sheaf theory and scheme theory are the language for expressing it faithfully, a language that is easy to use and natural, and aimed at explicitly describing the details of the inner structure of these geometric entities.

Classically, each affine variety has a corresponding coordinate ring that describes it algebraically by means of polynomial equations in an ambient space:

$$\text{affine variety} \Leftrightarrow \text{coordinate ring}$$

The fundamental idea of scheme theory is that this correspondence can be extended by associating each ring A with its *spectrum* Spec(A). We can see that the set of all the primes p of A gives rise to a collection of local rings A_p (germs). Vice versa, we want it to be possible to reconstruct A from this collection taken as a

---

[6]*Récoltes et Semailles*, pp. 28–30.

whole. This collection is actually the local reflection of an object – a sheaf – that is also topological in nature, so that A embodies the global aspect. The affine scheme Spec(A) results precisely from the synergy of topology (called Zariski topology) and the set of primes and the ranges of the corresponding local rings.

A scheme will thus admit a covering by affine schemes, that is, it is a topological space X and a structure sheaf O_x such that for every point of X there exists an open neighbourhood of the type Spec(A). The range now embodied by the structure sheaf follows and faithfully reflects the shape of the space underlying the scheme.

One advantage of this definition of *shape* consists mainly in the fact that it intrinsically describes geometric entities, schematically speaking, as a network of primary entities, omitting any reference to an ambient space. A further advantage of the scheme concept is its relative versatility, which makes it possible to conceive a scheme defined by a morphism based on what can even be a *family* of schemes.

A morphism of schemes X → S is simply a continuous application of the underlying spaces compatible with the structure sheafs. If S=Spec(A), such a scheme on S is equivalent to the fact that the structure sheaf O_X is a sheaf of A-algebras. For example, every scheme X can be considered as a scheme over S=Spec(Z).

Further, there exists a fibered product $X \times_s S' \to S'$ for schemes X → S and S' → S that effects the base change from S to S'. This product corresponds to the operation of extension or reduction of the scalars of the hypothetical equations for X. For example, every scheme is reduced modulo a prime number $p \in \mathbf{Z}$ by means of the product with S'=Spec($\mathbf{Z}/p$), producing in this way a family of schemes corresponding to the reduction modulo p of its hypothetical equations. Further, the product of X with S=Spec($\mathbf{C}$) produces a scheme in zero characteristic (an analytical space corresponding to the prime $p=\infty$). By isolating properties of "good behaviour" of the family by means of the concept of flat morphism, and rediscovering the concept of compactness by means of proper morphism, it is also possible to develop concepts of a differential nature in a purely algebraic context via the concept of smooth morphism.

These considerations led Grothendieck to develop systematically an algebraic geometry *relative* to the basis that makes it possible to "join together the various geometries associated with the various prime numbers".[7]

Seen in this way, a point of a scheme over a base will be simply a morphism of the base towards the scheme, and scheme may fail to have any point, that is, that it has points only when its base is changed.

An S-point of a scheme X→S is a morphism S→X that leaves S fixed. If k is a field S=Spec(k) it reduces topologically to a true point and the schemes of finite type over k, with their relative points, play the role of the new algebraic varieties, making it possible to visualise infinitesimal concepts by means of nilpoint elements. For example, the morphisms from Spec(k[$\varepsilon/\varepsilon^2$]) to a scheme X correspond to S-points of X over S=Spec(k) together with their tangent vectors.

---

[7] *Récoltes et Semailles*, p. 33.

In this sense the scheme $X \rightarrow S$ itself can be seen as a collection of fibres (X_s) as the points $s \in S$ of the base vary, but also as the collection of all of its points relative to the base, that is, as all the schemes $T \rightarrow S$ and morphisms $T \rightarrow X$ that leave S fixed. This vision of a scheme leads to the concept of representability that makes it possible to construct schemes by representing them by means of their hypothetical (relative) points.

Just as the concept of scheme constitutes a broadening of the concept of algebraic variety, the concept of topos constitutes a metamorphosis of the concept of topological space.[8] The *étale* topos and the *crystalline* topos associated with a scheme constitute the fundamental step for visualising the structure, that is, for the constructing the *cohomology invariants* of the scheme. With the concept of site already developed in 1958 – "the most fertile year of my whole life"[9] – Grothendieck similarly developed a relative topology in which some morphisms serve the role of open sets. The topos corresponding to such a site makes the arithmetic nature of the schemes completely clear. To put it briefly,

$$\text{scheme} \Rightarrow \text{topos} \Rightarrow \text{cohomology.}$$

> ...consider the set of all sheaves on a given topological space or, if you like, the prodigious arsenal of all the "meter sticks" that measure it. We consider this "set" or "arsenal" as equipped with its most evident structure, the way it appears so to speak "right in front of your nose"; that is what we call the structure of a "category"... From here on, this kind of "measuring superstructure" called the "category of sheaves" will be taken as "incarnating" what is most essential to that space. ...We can by now "forget" the initial space, keep and use the category (or arsenal) associated to it, which will be considered to be the most adequate incarnation of the topological (or spatial) structure that we intend to express.
>
> As often happens in mathematics, we have succeeded here (thanks to the crucial idea of sheafs and cohomological measuring stick) to express a given notion (that of a certain space) in terms of another (that of category). As always, the discovery of this kind of translation of one notion (which expresses a certain kind of situation) into the terms of another (corresponding to another kind of situation) enriches our understanding of both of them through the unexpected confluence of specific intuitions in relationship to each other. Thus, a "topological" situation (incarnated in the given space) or, if you will, the incarnated "continuum" of the space is translated or expressed by the structure of the category, which is "algebraic".[10]

According to Grothendieck, a *cohomology theory* naturally follows from the six operations associated to the category derived from the topos.

Grothendieck's *six operations* are functors between derived categories. They are the derived tensor product $\overset{L}{\otimes}$, the **R**Hom (which produces the values of Ext^i) and, for any scheme morphism f: $X \rightarrow S$, the direct image functors Rf_* and $Rf_!$ and the

---

[8]*Récoltes et Semailles*, p. 40.

[9]*Récoltes et Semailles*, p. 24.

[10]*Récoltes et Semailles*, pp. 38–39.

inverse image functors Lf_* and $Rf^!$. A theory of relative duality is expressed here by the adjunction between $Rf^!$ and $Rf_!$.

Grothendieck associates each geometry of characteristic p to an ℓ-adic cohomology corresponding to every prime $\ell \neq p$ by means of the étale topos, and a crystalline cohomology by means of the crystalline topos.

This arsenal of structures and operations is supposed to arrive at the same result. "It is in order to arrive and express this intuition of kinship between different cohomology theories that I have formulated the notion of motive associated to an algebraic variety".[11] This theme suggests that there is a common *motive* underlying the multitude of possible cohomology theories.

Grothendieck went on to suggest new conjectures, enhancing the unifying vision of the new geometry, the so-called "standard conjectures" that point to and predict the laws of a new *yoga* mediating between *form* and *structure*. While the Weil conjectures predicted the existence of a cohomology called, naturally enough, the Weil cohomology, later constructed by Grothendieck by means of the topos étale, that is, a structure associated to the form capable of grasping both the geometric and the arithmetic aspects, in the context of the dawning abstract (algebraic) geometry described above, Grothendieck's standard conjectures predict the existence of a motivic cohomology capable of synthesising in a single "invariant of the form" all of the structures that can be associated to it. Their formulation – obtained independently by Bombieri as well – appeared in a brief paper entitled "Standard conjectures on algebraic cycles" in the proceedings of the 1968 colloquium on algebraic geometry that took place in Bombay (Tata Institute of Fundamental Research, Mumbai).

The geometric construction of Grothendieck's motives is performed through the algebraic cycles that had already been introduced by Severi in the 1930s and then studied by Chow in the 1950s; these cycles are formal linear combinations of subvarieties and the correspondences from X to Y are defined by means of the cycles on the product $X \times Y$.

For a Weil cohomology $X \to H^*_\ell(X)$ there is a cycle map $Z^J(X) \to H_{\ell}^{2J}(X)$ which associates a cohomology class to every *algebraic cycle* of codimension j on X. The algebraic part of $H_{\ell}^{2*}(X)$ is that generated by classes of algebraic cycles. By means of Künneth's formula, we can also consider

$$H^*_\ell(X \times Y) = H^*_\ell(X) \otimes H^*_\ell(Y) = \mathrm{Hom}\,(H^*_\ell(X),\, H^*_\ell(X))$$

since $H^*_\ell(-)$ are vector spaces of finite dimension. The principle suggested by this identification is that cohomology operators of an algebraic kind have to be defined algebraically by means of a class associated to a cycle on the product, and thus by a correspondence.

---

[11]*Récoltes et Semailles*, p. 46.

The two *standard conjectures* can be briefly summarised like this: the first, called the *Lefschetz standard conjecture*, states that a given operator $\Lambda : H_\ell^*(X) \to H_\ell^*(X)$ which is the quasi-inverse of the Lefschetz operator L is induced by an algebraic cycle, that is, that the operator induced – by iteration – from the Lefschetz operator restricted to the algebraic part is an isomorphism. The second conjecture, called the *Hodge standard conjecture*, states that a given definite bilinear form on the primitive algebraic cohomology class is positive definite.

One simple consequence of the standard conjectures is the validity of Riemann geometric hypothesis as stated in the famous Weil conjectures, as well as the coincidence of the cohomological and numerical equivalence for algebraic cycles: an open question even in zero characteristic.

This mediating *yoga* based on the concept of *motive* and the corresponding theory of motives should provide the most refined structures associated with forms like invariants:

$$\text{form} \Rightarrow \text{motive} \Rightarrow \text{structure.}$$

Just as a musical motive has various thematic incarnations, so the motive can have various incarnations, or avatars, such that the familial structures of the (cohomological) invariants of the forms will be "simply the faithful reflection of properties and structures internal to the motive".[12]

The first congress entirely dedicated to motives took place in Seattle in 1991. Noteworthy advances in this area were achieved by Vladimir Voevodsky – winner of the Fields Medal in 2002 – who constructed a triangulated category of motives, by using methods from algebraic homotopy that had also been partly presaged by Grothendieck as "motivic homotopy types".[13] Voevodsky's construction makes it possible to obtain an "incarnation" of the motivic cohomology but it does not, however, find a solution to the standard conjectures, which are still today – along with the Hodge conjecture – the fundamental open question in modern algebraic geometry.

To close, Grothendieck and Einstein, through a "mutation of the conception that we have of space, in a mathematical sense on one hand and a physical sense on the other",[14] and an innovation in the way we look at the world via a unifying vision drawn from mathematics on the one hand and physics on the other, have turned out to be the mathematician and the physicist who revolutionised scientific thought through the concept of relativity.

---

[12]*Récoltes et Semailles*, p. 46.

[13]*Récoltes et Semailles*, p. 47.

[14]*Récoltes et Semailles*, p. 59.

Bibliography

All biographical information, bibliographies and more are available at the URL http://www. grothendieckcircle.org. A complete bibliography of Grothendieck's writings can also be found in the first volume of the *Grothendieck Festschrift* published by Birkhäuser in 1990. Here we note only a few essential sources used in this present work:

P. Cartier, "A mad day's work: from Grothendieck to Connes and Kontsevich, the evolution of the concepts of space and symmetry", *Bulletin of the American Mathematical Society,* vol. 38, no. 4 (2001), pp. 389–408.

J. Dieudonné, "De l'analyse fonctionnelle aux fondements de la géométrie algébrique", in *Grothendieck Festschrift*, Birkhäuser, Boston, 1990.

J. Giraud, "Une entrevue avec Jean Giraud, à propos d'Alexander Grothendieck", *Le journal de maths*, vol. 1 (1994), no. 1, pp. 63–65.

A. Grothendieck, *Récoltes et Semailles*, Montpellier, 1985-1986.

A. Grothendieck, "The responsibility of the scientist today", *Queen's Papers in Pure & Appl. Math.,* vol. 27, Kingston, 1971.

J.-P. Serre, *Grothendieck-Serre Correspondence*, AMS-SMF, 2003.

Gian-Carlo Rota

Mathematician and Philosopher

Domenico Senato

The first lesson I learned from Gian-Carlo Rota is expressed quite effectively by the three lines quoted in the preface to *Discrete Thoughts* of the poem by Antonio Machado, "Meditaciones del Quijote de José Ortega y Gasset":

> *Se miente más de la cuenta*
> *por falta de fantasia :*
> *también la verdad se inventa.*
> (The reason people so often lie is that they lack imagination: they don't realize that truth, too, is a matter of invention.)

Lies are told when fantasy is lacking, and we don't understand that even truth is invented. This idea permeated the intellectual and scientific life of Gian-Carlo Rota who, teaching and exploring mathematics and philosophy as a non-conformist, courageously and energetically re-examined current thinking of the time, revealing new, fascinating scenarios and touching profound levels of consciousness.

Rota was born in Vigevano on 27 April 1932 into a family seeped in culture. His father, an engineer and architect, owned an immense collection of books that included, in addition to works on architecture and engineering (today housed in the History Library of Vigevano's Technical Offices) volumes dedicated to mathematics, art, literature and philosophy. As a teenager Gian-Carlo sated his voracious curiosity with the books from his father's library, which quite soon led him not only to become interested in mathematics and philosophy, but to prepare himself unawares for the use of a computer, studying typing from a manual found among the other books. I still remember how impressed I was the first time I saw him at work in his home in Boston: his gaze glued to the monitor as he wrote a document in TEX, typing at an amazing speed, all ten fingers flying.

In addition to his father, young Gian-Carlo was deeply influenced by his aunt, Rosetta Rota, a mathematician educated in Rome under the guidance of Vito Volterra, later collaborator with the group of physicists in Via Panisperna, and wife of Ennio Flaiano. The well-known writer and set designer fascinated Rota. I have fond memories – from a stay in Rome in the summer of 1990 – of long evening walks that ended late at night, after having wandered through the streets in the centre of the city and visited all the sites of Flaiano's scenes and characters.

Gian-Carlo Rota

Gian-Carlo thus came to take part in the lively essences of Flaiano's condensed narratives, and was led to translate them into English, the language in which he felt most at home. Herbert S. Wilf, winner of the American Mathematical Society's Steele Prize in 1988, wrote, "Gian-Carlo Rota's ability to express himself in English, as opposed to his native Italian, was matchless. Listening to him we heard the Italian origins in his intonations and pronunciation, but he was rapier sharp in his use of English, and was never at a loss for exactly the right word. His sentences, both written and spoken, prepared and impromptu, were perfectly formed and featured a rainfall of extremely precise adjectives, colloquialisms, and so forth". Unfortunately, Rota's translations of Flaiano have never been published.

Rota began his studies in his home town, where, from 1939 to 1945 he attended middle school on a irregular basis because of the war and the vicissitudes of his family, which were the inspiration for his sister Ester's story "Orange sur le lac", published in France in 1995. In 1947, at 15-years old, Gian-Carlo followed his family to Equador, where his father moved in order to carry out his professional activities. In Quito Rota attended the American School; at 18 he moved to the United States and enrolled at Princeton, which was home in those years to some of the most brilliant mathematical minds, including Hermann Weyl, Kurt Gödel, Emil Artin, Solomon Lefschetz and Alonzo Church. The senior advisor for his thesis was William Feller. Rota paints a lively portrait in his essay "Fine Hall in its Golden Age" in the book *Discrete Thoughts*. About Feller, he wrote, "During a Feller lecture the hearer was made to feel privy to some wondrous secret, one that often vanished by magic as he walked out of the classroom at the end of the period". It is curious to note that a thoroughly similar sensation was often felt when one left the classroom after one of Rota's lectures.

At Princeton Gian-Carlo attended the classes in philosophy taught by Artur Szathmary and John Rawls, which led him to the study of phenomenology. Activities of philosophy absorbed a good part of his energy, and starting in 1972 he also held the chair in philosophy at MIT. An in-depth examination of this aspect of his intellectual activity is found in the well-documented book *La stella e l'intero* by Fabrizio Palombi, one of Rota's students and his collaborator in numerous works of philosophy. Here we will only mention two themes: the polemics with the analytical philosophers and the broadening of Husserl's concept of *Fundierung*, one of the cornerstones of his philosophical thinking.

The first criticism that Rota directed towards the analytical philosophers concerned the loss of speculative independence. The search for objectivity and rigour led many philosophers to use, in their investigations, axiomatic methods analogous to those used in mathematics, forgetting that mathematical results, even though checked and stated by an axiomatic method, cannot be achieved solely through their use. Rota maintained that confounding mathematics with axiomatics is like confounding Vivaldi's music with Baroque counterpoint. Traditional philosophical thought is quite distinct from mathematical thought. The only field in which a program of mathematicisation was successful was logic, and it was in any case for this reason that logic is today considered to be a branch of mathematics on a par with probability or algebra. According to Rota, many twentieth-century philosophers had submitted to the dictatorship of the incontrovertible, taking refuge in a slavish imitation of mathematics, considering the incapacity of giving definitive answers to be a failure of the philosophy of the past. At the base of this attitude lies a faulty judgment according to which the concepts – in order to make sense – have to be precisely defined. Even Wittgenstein was a prisoner of this way of thinking, later revising his youthful positions. Naturally, Rota is not against rigour, but he is opposed to the idea that the kind of rigour proposed by mathematics is the only one, and that philosophy should simply imitate it. In reality, even behind the fascinating steps forward in mathematics are found procedures analogous to those that give rise to thought, and Rota imagined that concepts that are today considered vague – such as motivation and purpose – could soon be formalised and accepted as constitutive elements of a new logic, in which they will be accorded a status, along with the notions of theorems or axioms, formalised over time.

In the concept of *Fundierung*, Rota identified one of the ideas capable of putting formal logic on equal footing with classic connectives, and perhaps capable of altering and enriching the structure of logic more than even Husserl himself had ever hoped. Rota – in keeping with his beliefs – never gave a definition of *Fundierung*, because there are no canons of definition in philosophy, but he clarified its meaning by a procedure of eiditic variations. For example, he examined the process of reading a text. The reading can take place as a physical procedure, if it is limited to merely factual observations. However, what matters most in reading is not the text but rather its meaning, and thus it is necessary to distinguish between the text and the meaning of the text. This is confirmed by the simple observation that the same meaning can be gleaned from the reading of different texts; so, the relationship between a text and its meaning is called *Fundierung*. Rota maintained that this relationship is constituted of two terms: *function* and *facticity*. The meaning of the text is a function correlated to the text by a relationship of *Fundierung* and, as such, cannot be traced back to questions of a physiological nature. According to Rota, sciences such as artificial intelligence which ignore difficulties of this sort are destined to fail. The distinction between function and facticity, which are made evident in the examples, becomes more difficult to delineate in the study of mental and psychological phenomena. Here, Rota suggested, is where a carefully compiled catalogue of *Fundierung* relations could turn out to be quite useful.

In 1954, Rota met Jacob T. Schwartz at the seminar in functional analysis organised at Yale by Nelson Dunford, and became his first doctoral student. Two years later he earned his Ph. D. with a thesis entitled "Extension Theory of Differential Operator I"; between 1958 and 1961, he published a series of papers in which he developed the idea of *theory of Reynolds operators*. Reynolds operators can be thought of as generalisations of the conditional mean operators and are a formidable tool for the unified treatment of ergodic theorems or *martingale* convergence theorems. Gian-Carlo's interest shifted to ergodic theory, which was at that time littered with difficult and sporadic combinatorial problems. Right away Rota sensed combinatorics' potential to develop into a mature and important field of mathematics. A few years later he would describe his impressions of that period, saying that rarely has a branch of mathematics, with perhaps the exception of number theory, been so rich in relevant problems and so poor in general ideas adequate for addressing them. On the other hand, every time the apparatus of technical instruments began to weigh heavily on the quality and the readability of the results in a mathematical subject, Rota, by shifting his point of view, was able to open a new and broader horizon for research.

In his essay entitled "Combinatorics, Representation Theory and Invariant Theory: the Story of a Ménage à Trois" in *Indiscrete Thoughts*, Rota separated the mathematicians into two broad categories: problem solvers and theorists. Even though he admitted that in general mathematicians possess a little of both qualities, he said that it is not unusual to find extreme cases in each of the two classes. Alfred Young, for example, was mostly a problem solver, while Hermann Grassmann was most certainly a theorist. His most important contribution was the definition of exterior algebra, which he developed and refined throughout his life, anticipating the calculus of exterior differential forms that would be developed by Élie Cartan in the next century. For a problem solver, what counts is getting to the heart of a problem, even better one that is considered to be unsolvable, no matter how complex, difficult, or difficult to interpret it may be. What matters is to have found it and be sure of its correctness. A problem solver is essentially a conservative, so the conceptual background to which he refers has to remain unchanged over time; new theories or generalisations are regarded with suspicion. For a theorist, on the other hand, the greatest contribution to mathematics is not the solution to a problem, but rather the elaboration of a new theory in which the problem finds a natural solution. The theorist is a revolutionary, convinced that his own theories will still be vital when the fashionable problems of the moment have made it clear how bulky all of the by-now obsolete techniques used to solve them were. There is not doubt that Rota felt closer to the theorists than to the problem solvers. In the preface to Joseph Kung's *A source book in matroid theory*, Rota proposed a criterion for distinguishing between the three ages of a mathematical subject. The oldest ones are covered with awards and honours, whose most important problems were solved long ago and whose applications are a copious harvest for engineers and entrepreneurs: their ponderous treatises are stacked on dusty shelves in library basements, until the day when a new generation yet unborn shall rediscover with awe that paradise lost. To give an idea of the middle aged subjects, we need only browse the corridors of the *Ivy League universities* or the *Institute for Advanced Study*, their high priests haughtily refusing the fabulous offers from anxious provincial universities, knowing all the while that burden of technicality has already reached a critical mass that threatened to submerge their theorems in the dust of oblivion. Finally, the youngest subjects. They are born thanks to individuals who are a little eccentric, who energetically hack away at a mountain of impossible problems, naively stuttering out the first words of what will become a new language. The infancy is over with the first *Bourbaki seminar*. Rota showed an extraordinary capacity for transforming disjointed heaps of combinatorial problems that characterised the mathematical panorama of the 1960s, into a young subject based on the solid foundations of algebra: algebraic combinatorics.

In the years from 1959 to 1965, Rota was first assistant professor and then associate professor at MIT, where he would return after a 2-year parenthesis at Rockefeller University. At MIT he met Norbert Wiener and John Nash. Rota would never again abandon MIT, Cambridge and the city of Boston.

In 1964 he published "On the Foundation of Combinatorial Theory I. Theory of Möbius Functions", the first of ten papers published between 1964 and 1992 which profoundly influenced the directions taken by research in contemporary combinatorial theories. For this first paper, which marked the beginning of modern algebraic combinatorics, he was awarded the 1988 *Steele Prize* from the *American Mathematical Society*, which had this to say:

Only 25 years ago the subject of combinatorics was regarded with disdain by "mainstream" mathematicians, who considered it as little more than a bag of ad hoc tricks. Now, however, the new subject of "algebraic combinatorics" is a highly active and universally accepted discipline. Two of its most prominent features are its unifying techniques which bring together a host of previously disparate topics, and its deep connections with other branches of mathematics, such as algebraic topology, algebraic geometry, commutative algebra, and representation theory. The single paper most responsible for bringing on this revolution is the paper of Rota cited above. It showed how the theory of Möbius functions of a partially ordered set, as developed earlier by L. Weisner, P. Hall, and others, could be used to unify and generalize a wide selection of combinatorial results. Moreover, it hinted at connections with algebra, topology, and geometry which were later to be extensively developed by Rota and his followers. Today the theory of Möbius functions occupies a central position within algebraic combinatorics and has found many applications outside combinatorics. Perhaps more importantly, Rota's paper has inspired many mathematicians to develop systematic techniques for solving combinatorial problems and to apply them to problems outside combinatorics.

This first paper, like the others in the series on the foundations of combinatorics, produced a abundant crop of further research. For example, Rota's intuition that the Möbius function of a lattice can be interpreted in different ways as a Euler characteristic, paved the way for the study in innumerable problems related to topology, leading to the birth of a new subject: topological combinatorics. Today this theory has achieved an elevated degree of conceptual refinement. Again, the relationships between the Möbius function and geometric lattices have revitalised matroid theory – objects based on a generalisation of the concept of linear independence – and revealed the profound connections of these objects with topology and algebraic geometry.

Rota's return to MIT in 1967 marks the beginning of the Cambridge school of combinatorics. Gian-Carlo gathered around him those who would soon become some of the major leading figures in the skyrocketing growth of combinatorics. The seminars that took place weekly at MIT hosted figures of great renown, such as Marcel-Paul Schützenberger, and were attended by scholars of the calibre of Danny Kleitman, Henry Crapo, Jay Goldman and by graduate students or junior faculty members whose names would soon become famous, such as Richard Stanley, Peter Doubilet and Curtis Green. During those same years, Rota was intensely engaged in activities of publication as well, founding the *Journal of Combinatorial Theory* and *Advances in Mathematics*, two journals which would rapidly achieve great prestige internationally. Edwin F. Beschler, at that time the acquisitions editor for mathematics at Academic Press, wrote:

> It was Gian-Carlo's particular genius that he could transform an intractable set of dynamics sheerly by force of his ability to recognize superior work and his willingness to "break the rules" in the interests of publishing it expeditiously, thus furthering mathematics. He was a communicator of the highest degree, and he believed in the power of the written word and the necessity – even to proliferation – of publishing thoughts, ideas, and information.

In addition to the two journals mentioned, Rota was one of the promoters of the birth of the *Journal of Functional Analysis* and founder in 1979 of *Advances in Applied Mathematics*, which, in just a few years, achieved a prestige of its sister journal. His activities in promoting publications were unending. Among the innumerable initiatives, also deserving of mention are the series *Contemporary Mathematicians*, published by Birkhäuser, and *The Encyclopaedia of Mathematics*, published by Cambridge University Press, comprising more than 80 volumes.

The year 1964 was crucial for Gian-Carlo, not only because of the publication of "On the Foundations of Combinatorial Theory" but also because it was the year he met Stanislaw Ulam, one of the greatest exponents of the Polish school of mathematics, and a collaborator of von Neumann's. Between Ulam and Rota sprang up an intense intellectual relationship as well as a friendship, which led the Polish mathematician to suggest Gian-Carlo as consultant in the management of the celebrated *Los Alamos Scientific Laboratory*, a collaboration that Rota would carry on continuously until his death. Ulam wrote,

Rota impressed me by his knowledge of some half-forgotten fields, the work of Sylvester, Cayley and others on classical invariant theory, and by the way he managed to connect the work of Italian geometers to Grassmannian geometry and modernize much of this research which dates to the last century.

In truth, many of Rota's most elegant and profound contributions were born of his passion for the works in combinatorics by the mathematicians of the nineteenth century. Conscious of mathematics' unhistoric nature, he conceived development as a complex pathway, with motions of reflux that show that progress is decisive only when it filters its heritage, reinterprets its roots and deepens its foundations. This is the case of the prodigious undertaking that he began in the early years of the 1970s, which led to the rebirth of the classic invariant theory. It is no coincidence that the first mathematicians to deal with theories of combinatorics were also "invariantists". Hammond, MacMahon and Petersen are known today for their work in combinatorics, but the motivation underlying their research was invariant theory. In a similar way, the names of Cayley, Clifford and Sylvester are firmly tied to the theory of invariants, but their contributions to combinatorics were quite significant. In his 1999 article entitled "Two Turning Points in Invariant Theory", Rota wrote,

The program of invariant theory, from Boole to our day, is precisely the translation of geometric facts into invariant algebraic functions expressed in terms of tensors. This program of translation of geometry into algebra was to be carried out in two steps. The first step consisted in decomposing tensor algebra into irreducible components under changes of coordinates. The second step consisted in devising an efficient notation for the expression of invariants for each irreducible component.

It was precisely Rota's search for an efficient notation that led Rota to follow in the footsteps of Gordan, Capelli and Young and their symbolic techniques. However, for Gian-Carlo, the symbolic method gave rise to more than only an efficient notation; in the same article, he goes on to say,

The hidden purpose of the symbolic method in invariant theory was not simply that of finding easy expression for invariants. A deeper faith was guiding this method. It was the expectation that the expression of invariants by the symbolic method would eventually guide us to single out the "relevant" or "important" invariants among an infinite variety.

The method worked out by Rota took as its point of departure an idea of Richard Feynman's. The physicist represented monomials of noncommutative algebra, substituting for each variable a pair of variables, the first of which indicated the original variable, while the second "marked" the place that the variable occupied in the noncommutative monomial. By means of this expedient, a pair of variables can be interpreted as a single variable that generates a commutative ring, and many problems of noncommutative algebra can be traced back to problems of commutative algebra. Rota understood that the same idea could be used to deal with problems in combinatorics that arose from invariant theory. He gave the name *letter-place algebra* to the algebra of pairs of variables constructed in this way.

Gian-Carlo told me about his last meeting with Feynman, at the inauguration of the first *Connection Machine* at the "Thinking Machines Corporation". He told

Feynman that he had used the idea of the pairs of variables, successfully, in many articles. The physicist immediately left the crowd of journalists that surrounded him, took Gian-Carlo aside, and confided to him with satisfaction that he considered the temporal ordering – that was the term he used for letter-place algebra – to be the best idea he'd ever had. He was convinced that it was even better than the Feynman integral. He then went on to explain to Rota another idea that he had never published, making a sketch of it that was no larger than a postage stamp. Gian-Carlo put it in his pocket, with the idea of pulling it out later. To his great disappointment, he later discovered he had lost it. He had asked himself ever since what Feynman's last idea had been.

Letter-place algebra also made it possible to construct the fundamental straightening algorithms used by Rota and his collaborators not only to reformulate in modern terms the classic themes of invariant theory of Hermann Weyl and the results in positive characteristic of Jun-ichi Igusa, but also provided a unifying approach for use in different contexts, such as the ordinary representation theory of the symmetry group and the representation theory of the general linear group and the symmetry group of the space of homogeneous tensors. Gian-Carlo believed that in order to understand the difference in style as well as in content between representation theory and invariant theory it was helpful to consider the analogous differences between probability theory and measure theory: "A functional analyst could spend a lifetime with measurable functions without ever suspecting the existence of the normal distribution. Similarly, an algebraist could spend a lifetime constructing representations of his favorite group without ever suspecting the existence of perpetuants".

Thanks in part to the contribution of Rota, representation theory today is a very active area in contemporary combinatorics. Central to it is the most explicit and efficient construction possible of the irreducible representations both of the classic groups as well as of the Coxeter groups and the algebras associated to them. The main combinatorial tools are symmetric functions and their generalisations, such as Schubert's polynomials and the various versions of Schensted's classic algorithm. In multilinear algebra, Rota's fundamental straightening algorithms can be considered as analogous to Schensted's combinatorial algorithm. With the introduction of supersymmetric variables, that is, substituting external algebras with suitable tensor products of fields, the straightening algorithms of letter-place algebras were made considerably more powerful. The union of commutative variables and noncommuntative variables was used for a long time by physicists and only later by mathematicians. The contributions of Rota, Brini, Grosshans, Stein and others enriched the tool kit with instruments that were highly effective. In particular, the concept of the polarisation of variables and the definition of an umbral operator led to the solution of numerous classic problems and an extraordinary simplification in the representation theory of Lie algebras. The adjective "umbral" opened an important chapter in the scientific life of Gian-Carlo Rota. Alain Lascoux, one of Marcel-Paul Schützenberger's best known students, wrote that Rota thought of himself as a writer of epigraphs of the richness of the past and an advocate of the algebraic structures that make it possible to integrate that richness into contemporary research. The development and spread of umbral calculus, whose applications are rapidly being developed today, effectively confirm this judgment. Extensive use has been made of the so-called "umbral calculus" since the nineteenth century, even though it had no rigorous foundations. It was born of observations of analogies between various sequences p_n of polynomials and the sequence of powers x^n. For example, as x^n provides the number of applications between a set of n elements and a set of x elements, so the decreasing factorial sequence $(x)_n = x(x - 1)...(x - n + 1)$ provides the number of injective applications between the same sets. Thus the index n, in the polynomial sequence, can be considered as an "umbra", or shadow, of the exponent of x. In the nineteenth century many identities were first established using the trick of substituting the exponents with the index and then verifying the results a posteriori. This technique was developed by Rev. John Blissard in a series of papers starting in 1861. Blissard's calculus grew out of the symbolic methods for successive derivatives of a product with two or more factors invented by Leibniz, which was later developed by Laplace, Vandermonde, Herschel, and enriched by the contributions of Cayley and Sylvester in the theory of forms. In 1940 Eric Temple Bell tried to provide a rigorous foundation for these techniques, although he was not completely successful. In 1958, Riordan (in his book *An Introduction to Combinatorial Analysis*, which can be considered the first modern text in combinatorics) made wide use of umbral techniques, without providing any proofs of the method's correctness. Just 6 years later, in 1964, Rota published "The Number of Partitions of a Set", where he revealed "umbral magic" that made it possible to obtain identities by substituting indexes for exponents, defining the linear functional that legitimises the method.

This article would pave the way to an elegant theory, set out in the articles "Foundation III" and "Foundation VIII", which would give rise to an incredibly vast number of applications in different fields of mathematics. In 1978 Rota and Roman provided a definitive formal structure for the whole subject in the language of Hopf algebras. Sixteen years later, Rota returned to umbral calculus, carrying out Bell's dream of providing it with a solid base in algebra while maintaining as far as possible the spirit of its founders Sylvester and Blissard. The new approach opened innovative perspectives and restored to it the intuitive power and simplicity which the translation into the language of Hopf algebras had partially obscured. The most recent developments confirm what a powerful tool for calculation and simplification it is in contexts as diverse and significant as *wavelet theory* and probability.

At the end of his life, Gian-Carlo came full circle, sowing new seeds in the field in which he had first been Feller's student at Princeton. He died, still in his intellectual prime, in his house in Cambridge, in April 1999. MIT dedicated a hall to his memory, the Gian-Carlo Rota Reading Room, which contains an ample collection of books related to his intellectual itinerary, a testament to the scope and depth of his thinking.

Stephen Smale

Mathematics and Civil Protest

Angelo Guerraggio

Moscow, August 1966: in the course of the International Congress of Mathematicians, or ICM, held every 4 years, the Fields Medal is awarded. Whether or not it is actually the equivalent of the Nobel prize, it is in any case the most prestigious international prize. The year in Moscow the winners are the English Michael Atiyah, the American Paul Cohen, the French Alexander Grothendieck, and the American Steve Smale.

It was Smale who managed, with a few fine strokes, to turn the mathematical–scientific event into a political event as well, one with which the newspapers (beginning with the *New York Times*) were forced to cover.

In 1966, Smale was 36-year old, but – as his winning the Fields Medal testifies – he was naturally no freshman as a mathematician. He had studied at the University of Michigan, where he had chosen as the advisor for his doctorate thesis, a mathematician – Raoul Bott – who had taught a course in algebraic topology (building on some ideas of J. P. Serre). Bott was not particularly famous, but he had set Smale on the trail of a "good problem". The thesis dealt with the regular closed curves on Riemann manifolds, with the aim of classifying them in the absence of a regular homotopy, and generalising the results obtained in the plane by H. Whitney in 1937.

After his thesis, Smale began to travel, partly because his credentials weren't excellent, and it was difficult to get a foot in the door of the academic world. Saunders MacLane, for example, had a good impression of him, but still harboured doubts. At a congress in Mexico City, Smale had the good fortune to meet John Milnor – he sat in on his seminars on the new differential topology – and especially René Thom, who "introduced" Smale to tranversality. He also came to know Marston Morse personally, and Morse theory was to become of the main "ingredients" in Smale's mathematics. But in this case, this personal connection didn't lead to anything great. However, Smale's well-known (and counter-intuitive) result regarding the eversion of the sphere, appeared in 1959 in the *Transactions of the American Mathematical Society* as "A Classification of Immersions of the Two-Sphere".

C. Bartocci et al. (eds.), *Mathematical Lives*,
DOI 10.1007/978-3-642-13606-1_27, © Springer-Verlag Berlin Heidelberg 2011

Stephen Smale

Smale then went to Brazil, and while on the beaches of Rio he successfully solved Poincaré's conjecture – the most famous problem then still unsolved in topology – setting the stage for his winning the Fields Medal. For a certain period of time, he even thought he might win it in 1962, during the Stockholm congress, but he had to reconcile himself with its being awarded to Milnor (and to Lars Hörmander). The generalised conjecture of Poincaré states that an n-dimensional manifold, closed and homotopically equivalent to an n-dimensional sphere, is homeomorphic to it. Smale proved this for $n \geqslant 5$, disproving along the way the idea that an increase in the dimension creates greater difficulties; the higher dimensions can even turn out to be easier to deal with, because there is more "space" to move around in. The positive answer to Poincaré's conjecture, for $n=4$, wouldn't be given until 1982, thanks to Steve Friedman. That relative to the case of $n=3$ was more recently given by the Russian mathematician Grigorij Perelman. The generalised conjecture was actually proven for $n \geqslant 5$ independently and just a few months later by John Stallings and by Christeroper Zeeman; in fact, some surveys (including Morris Kline's history) rank the contributions of Smale, Stallings and Zeeman equally. Smale wasn't very pleased about this, and already smarting from some earlier experiences (including missing the 1962 Fields Medal award by a hair), struck back with an obstinate vindication of priority in 1989, on the occasion of the annual meeting of the American Mathematical Society. The title of his talk –

in itself anomalous and provocative – is "The Story of the Higher Dimensional Poincaré Conjecture. (What Actually Happened on the Beaches of Rio)".

In 1966 in Moscow, Smale was already a well-known figure from a political point of view. He had been brought up by his father with a mentality that was very anticlerical and leftist. At the University of Michigan, he embraced Marxist ideas and joined the Communist Party, attending meetings of its youth organisation – the Labor Youth League – and also acting as its representative at a Peace Festival in East Berlin. His political activities were genuinely militant – not to mention secret – and would lead him to neglect his studies to some extent. He only began to speak openly about this in the 1980s:

> But still to accept the Communist Party?
> Consider my frame of reference at that time. I was sufficiently skeptical of the country's institutions to the point that I couldn't accept the negative reports about the Soviet Union. I so believed in the goal of a utopian society that brutal means to achieve it could be justified. I was unsure of myself on social ground, and the developing social network of leftists around me gave me security.

After his militancy at university, Smale – as we have seen – returned to his mathematical studies in earnest. His political sympathies were, however, still decidedly leftist. This is why he didn't hesitate to side with Castro's revolution (even getting in touch with an organisation named "Fair Play for Cuba", with which Lee Harvey Oswald, Kennedy's assassin, was also involved for a period of time).

But he returned to more active politics when he accepted a position as associate professor at Berkeley, following stints in Chicago, and at Princeton's Institute for Advanced Study (1960). He then transferred to Columbia University, only to return to Berkeley in 1964 as full professor. The University's administration was trying out a more strict approach on campus towards both the freedoms and political activities of students. As a result, the Free Speech Movement was born on campus to counter the new policies. This movement's golden period lasted a mere 3 months – from September to December 1964 – but those 3 months, if they didn't exactly change the world, certainly changed that particular generation, along with the habits and culture of Western society. The confrontation between the movement and the president and administration of the university was violent. The number of sit-ins grew. Joan Baez sang "We Shall Overcome". But there was also the police, and their attitude towards the situation was not exactly comforting. Smale came out strongly on the side of the students, without any of the doubts that his so-called liberal colleagues expressed. He gave them concrete help and support. And in the end, they won. A small group of students was able to defeat the powerful university administration, overturning the roles traditional assigned to students, faculty, administration. It all started in Berkeley, then it spread to Europe. Then there was 1968.

But now we are still in the winter of 1964, and in American cities in 1964 the protest movement against the war in Vietnam was practically non-existent. Once again, it all began in the universities, and Smale was on the front line. As early as 1965 we find him side by side with the students protesting against militarism in the United States. He was one of six faculty members who participated in a teach-in.

As chair of Berkeley's Political Affairs Committee, he obtained approval for a motion condemning the air attacks on Vietnam "which notably increased the risk of a world war". He then founded, along with his wife Clara, a student and a colleague, the Vietnam Day Committee, whose purpose was to organise a day of protest against the war in Vietnam. The event turned out to be a huge success. The 24 h of the sit-in were extended to 30, in order to give everyone a chance to speak. Even the noted paediatrician Benjamin Spock took part. The newspaper only gave minimal coverage to the event, but Secretary of State Dean Rusk began to be worried: "I continue to hear and see nonsense about the nature of the struggle. I sometimes wonder at the gullibility of educated men and the stubborn disregard of plain facts by men who are supposed to be helping our young to learn". There appeared the first instances of civil disobedience, and the first invitations to desert the military. Smale didn't hesitate to recall how the Germans remained ignorant for a long time about the atrocities committed by the Nazis during the second world war. Then there was the march in Oakland. It was now October 1965, and Americans had by this time become aware of the horrors and futility of the war. In April 1967 there was the great march in Washington. Robert Kennedy and Martin Luther King would make their voices heard. In the Pentagon, McNamara began to think that the war was by this time a lost cause, and President Johnson "found" him a new job as president of the World Bank. But even Johnson was forced to surrender to the truth, and soon announced to the nation that he would not campaign for another term as president.

Now we can get back to Moscow and the summer of 1966. Smale was in Europe, on his way to the Soviet capital, and while in Paris he took part in one of the "Six Hour" events in support of Vietnam organised by Laurent Schwartz. His speech – that of an American citizen protesting against his own country's militarism – was particularly looked forward to, and actually turned out to be quite moving. In his autobiography, Schwartz recalled that the most emotional moment of the meeting was the handshake between the American Steve Smale and the Vietnamese Mai Van Bo. Then, in the company of René Thom, he headed for a conference in Geneva. It was during that trip that Thom, a member of the Fields Medal committee, told him about his "victory". From Geneva, Smale went to Greece for a short vacation with his family. It was here, at the airport, on 15 August, the ICM's opening day, that Smale was stopped for an irregularity in his passport. He risked missing everything (seeing Moscow, attending the ICM, receiving the Fields Medal during the opening ceremony). As it turns out, he was somehow permitted to leave, and arrived some hours late. Not having his ICM badge – he had not had time to register and pick it up – he was stopped again, and lost more time. When he was finally allowed in, the medals had already been awarded, and he was just in time to hear René Thom read the official motivation for the award: "If Smale's works perhaps do not possess the formal perfection of definitive work, it is because Smale is a pioneer who takes risks with a tranquil courage".

Smale also wanted to take advantage of the Moscow congress – in agreement with Schwartz and a few other mathematicians – to launch an appeal for signatures condemning American aggression and in support of Vietnam. But in 1966, the Soviet bureaucratic apparatus was strict, even when the contents of the appeal were in line with the politics of the communist bloc! One of prime objectives of hosting the ICM in Moscow – at a time when Moscow saw very few international events – was to show that the USSR was a friendly, peaceful country. In the meantime, Smale was approached by a Vietnamese journalist asking for an interview. He decided to organise a press conference, inviting the Vietnamese journalist as well as members of the Soviet and American press. This is where he made his statement:

> I believe that American military intervention in Vietnam is horrible and becomes more horrible everyday. I have great sympathy for the victims of this intervention, the Vietnamese people. However, in Moscow today, one cannot help but remember that it was only ten years ago that Russian troops were brutally intervening in Hungary and that many courageous Hungarians died fighting for their independence. Never could I see justification for military intervention, 10 years ago in Hungary or now in the much more dangerous and brutal American intervention in Vietnam. ... I feel I must add that what I have seen here in the discontent of the intellectuals on the Sinyavsky-Daniel trial and their lack of means of expressing this discontent, shows indeed a sad state of affairs. Even the most basic means of protest are lacking here. In all countries it is important to defend and expand the freedoms of speech and the press.

The organisation of a press conference by an American in Moscow in 1966 was considered to be a subversive act. But the Soviets could not take drastic steps, because they would lose face in the eyes of international public opinion. Smale was saying that Vietnam (and its communist allies) were right! Thus, they decided to limit the damage. They detained Smale in order to prevent his having other contacts – treating him with all the respect due to a diplomat, and extending many a privilege to him – but they sent him away from Moscow under pretext of a long sight-seeing tour, which only came to an end when Smale had had enough and demanded to be let out of accepting other similar "courtesies". The trip had been a long one: it was late night before he was able to return to his hotel. The morning after, at 7:00, a plane was waiting to fly him back to Athens.

But his troubles weren't over. Now it was the Americans turn. The *New York Times* came out with a front-page story about what had taken place in Moscow, including the press conference. The day after, the National Science Foundation (NSF) opened an investigation into Smale's case. As a precaution, all of his funding was suspended. The main accusations were of having attacked the government of the United States while abroad, having travelled to Europe using government funds intended for research – demanding that he prove that he actually had worked on mathematics over the summer – and having taken a non-American ship (French) to return home. Smale had to explain himself, guaranteeing in writing that he had worked on research even while camping, in a hotel, and on the ship. Thus, "On the S.S. France, for example, I discussed problems with top mathematicians and worked on mathematics in the lounge of the boat. (My best known work was done on the beaches of Rio de Janeiro, 1960!)". This is how the story of the beaches in Rio came to light. Smale was supported by colleagues and some members of

Congress. A compromise was proposed, but Smale refused it, declaring that the "NSF has dishonored itself". A compromise would have meant capitulating, which would have made it difficult – for other researchers – to choose to public disassociate themselves from "Johnson's brutal policies in Vietnam". It would take until the end of September, but in the end Smale's hard line was rewarded.

After Moscow, of course, Smale's life and mathematics carried on. Research in algebraic geometry, differential topology and the theory of h-cobordism were flanked by work in dynamic systems, which had already begun during his first stay in Brazil (again on the beaches?). Then – by now we have reached the 1970s – he developed an interest in mathematical economics. It was Nobel Prize laureate Gerard Debreu who, in Berkeley, involved him in the problem of general economic equilibrium. Smale's approach was completely original, and consisted in trying to make it into a dynamic system – the "mathematics of time" with the reintroduction of the classic hypothesis of differentiability, in place of that of convexity (which had been stated starting in the 1930s). He also dealt with vector optimisation and critical points. He then turned his attention to linear programming and analysis of algorithms. His most recent interests are their study, the calculability, and some problems of theoretical computer science. In 1998 he published *Complexity and Real Computation*.

The Moscow press conference meant that there was no celebration of his having received the Fields Medal in the United States, but he received his due in 1996 (30 years later) when he was awarded the National Medal of Science by then president Bill Clinton.

Today Smale is still active as a mathematician. He studies, writes, takes part in international congresses, speaking above all about themes related to human and artificial intelligence. He still collects minerals, and has been ranked as one of the first five collectors worldwide in his category. A brilliant mathematical mind and a passionate defender of civil rights, he has still been able to find time to appreciate nature.

Smale however no longer lives in Berkeley. Once he retired, he decided to accept one of the many offers that he received and became professor emeritus at a university in Hong Kong. A final (for now) provocative gesture by an old "sixties radical". The mathematician who succeeded in condemning, at a single stroke, two superpowers and two economic systems in Moscow, now lives in the postmodern hybrid of communist capitalism!

Michael F. Atiyah

Mathematics' Deep Reasons

Claudio Bartocci

The Age of Unification

How will the mathematics of the last 50 years appear to the eyes of future historians? As difficult as it is to make predictions that depend in large part on developments to come (Lakatos *docet*), we can hazard a guess that the second half of the twentieth century will probably be considered a period of extraordinary proliferation of new ideas, and at the same time, of the rediscovery of the fundamental unity of mathematics. In the first half of the century, marked by the triumphant of Hilbert's program culminating in the great undertaking of the Bourbaki group, there was a widespread tendency – we might say – towards specialization and therefore towards the parcelling out of mathematical knowledge: doubtless, it was this tendency that made possible the rapid development of disciplines such as general topology, group theory, mathematical logic, differential geometry, functional analysis, algebraic and differential topology, commutative algebra and algebraic geometry (even if this last was destined to change its face in later years). On the contrary, the second 50 years are characterized above all as an "era of unification, where borders are crossed over, techniques have been moved from one field into the other, and things have become hybridised to an enormous extent"[1]: cross-fertilization has prevailed as the dominant paradigm. There are many examples that lend support to this thesis: the admirable theoretical edifice constructed by Grothendieck; the development of research areas such as global analysis; the omnipresence of category theory methods; arithmetic geometry or Alain Connes's non-commutative geometry; the formidable results obtained by mathematicians such as Vladimir Arnol'd, Yuri I. Manin, Shing Tung Yau, Simon K. Donaldson, Vaughan Jones, Edward Witten, Richard Borcherds, Robert

The author warmly thanks Sir Michael Atiyah for reading a preliminary version of this article and for pointing out a few inaccuracies.

[1] M. F. Atiyah, "Mathematics in the twentieth Century", *Bulletin of the London Mathematical Society*, 34 (2002), pp. 1–15.

C. Bartocci et al. (eds.), *Mathematical Lives*,
DOI 10.1007/978-3-642-13606-1_28, © Springer-Verlag Berlin Heidelberg 2011

Langlands and Andrew Wiles. But perhaps more than anyone else, the person who demonstrated how fruitful it can be to delve into different disciplinary contexts in order to arrive at fundamental discoveries, to dig down to the roots of problems that are apparently distinct in order to identify the profound reason that unites them, is Michael Francis Atiyah, without a doubt one of the most prolific and influential mathematicians of the past century.

Michael F. Atiyah

From Algebraic Geometry to K-Theory

Born in London on 22 April 1929 to a Lebanese father and Scottish mother, Atiyah spent the years of his boyhood in Khartoum, Sudan and Cairo.[2] In 1945, at the end of World War II, his family moved to England; the young Michael was sent to Manchester Grammar School, because it was reputed to be the best school for mathematics in England. After having spent 2-years in the National Service, he enrolled at Trinity College in Cambridge: his first article, written in 1952 under the

---

[2]Atiyah himself provided detailed information about his life and his research activities in a series of talks, recorded in March of 1997, available (along with their transcripts) on the website http://peoplesarchives.com. See also M. F. Atiyah, *Siamo tutti matematici*, Di Renzo, Rome, 2007.

guidance of J. A. Todd, regarded a question of projective geometry. Having begun his doctoral studies, still at Cambridge, Atiyah chose as his supervisor one of the greatest English mathematicians, William V. D. Hodge, who in 1941 had published his famous treatise on harmonic integrals. Hodge directed him to the work of Chern on characteristic classes; Atiyah went to discover the new ideas – developed above all by the French school (André Weil, Henri Cartan, Jean-Pierre Serre) – which were emerging in those years in algebraic geometry.[3] He got acquainted with vector bundles, coherent sheaves and their cohomology, and became a voracious reader of the *Comptes rendus*. In 1955–1956 Atiyah spent a long period Institute of Advanced Study in Princeton, where he came to know Raoul Bott, Friedrich Hirzebruch and Isadore M. Singer and Serre, and would broaden his horizons in an atmosphere that was very stimulating intellectually.

Until 1959 the most part of Atiyah's works were in algebraic geometry: of particular note, for example, is the 1957 article on the classification of vector bundles on an elliptic curve. In later years – after his return to Europe – his interests would turn above all to topology. This reorientation was principally due to the influence of the German mathematician Friedrich Hirzebruch, organizer of the famous *Arbeitstagung* in Bonn, in which Atiyah was a faithful participant. Hirzebruch had extended (and reinterpreted in a topological key) the classic Riemann–Roch theorem to algebraic manifolds of higher dimensions and, making use of Thom's cobordism theory, had defined particular combinations of "characteristic numbers" that assume integer values not only for algebraic manifolds, but for all differentiable manifolds as well. Hirzebruch's results, united to Grothendieck's profound generalization of the theorem of Hirzebruch–Riemann–Roch and to Bott's theorem of periodicity,[4] are the ingredients at the base of topological K-theory, which Atiyah (in collaboration with Hirzebruch) set out between 1959 and 1962.[5] Atiyah himself recalls: ". . . I saw that by mixing all these things together you ended up with some interesting topological consequences, and because of that we then thought it would be useful to introduce the topological K-group as a formal apparatus in which to carry this out".[6]

K-theory – which can be thought of as a kind of generalized cohomology theory constructed beginning with the isomorphism classes of vector bundles – immediately showed itself to be a useful and versatile tool for tackling problems of various natures: one of the most remarkable example was the solution by Frank Adams in 1962 of the problem of the maximum number of nowhere vanishing and (pointwise) linearly independent tangent vector fields on an odd-dimensional sphere (by the

---

[3]Cf. Jean Dieudonné, *History of algebraic geometry*, Wadsworth, Monterey 1985, Chap. VIII.

[4]Bott's theorem of periodicity regards the homotopy groups of the groups U(n) as $n \to \infty$.

[5]For mathematical details not included here, see M. F. Atiyah, "K-theory past and present" in *Sitzungsberichte der Berliner Mathematischen Gesellschaft,* Berliner Mathematischen Gesellschaft, Berlin 2001, pp. 411–417 and M. F. Atiyah, "Papers on K-theory", in *Collected Works*, vol. 2, Oxford University Press, New York, 1988, pp. 1–3.

[6]M. F. Atiyah, http://www.peoplesarchives.com (Part 7, 36).

famous "hairy ball theorem", the $2n$-sphere has no non-vanishing vector field for $n \geq 1$).

At the International Congress of Mathematicians in Stockholm in 1962, Atiyah presented the numerous, and in large part unexpected, applications of K-theory, and in conclusion, mentioned the new interpretation which admits the notion of "symbol of an elliptic operator" into this theory. In the unifying framework of "K-theory" converged methods and ideas coming from algebraic geometry, topology and functional analysis: the "index theorem" – already clearly stated in the summer of 1962 but still awaiting proof – will be the culmination of a long series of researches.

The Index Theorem

Atiyah spent the years from 1957 to 1961 in Cambridge as a lecturer and tutorial fellow at Pembroke College, overburdened with many hours of teaching. In 1960 the topologist Henry Whitehead died at 55, and his chair in Oxford remained vacant: Atiyah applied for the chair unsuccessfully (the favorite was one of Whitehead's students, Graham Higman) and as makeshift solution, which in any case freed him from excessive teaching duties, he accepted a position as a reader. Less than 2 years later, with the death of Titchmarsh, the prestigious chair of Savilian Professor would become free, and Atiyah was called to hold it.[7]

In the attempt to extend results valid in the case of algebraic manifolds to differentiable manifolds, Hirzebruch had proven that a certain combination of characteristic classes – the so-called Â-genus – which is in general a rational number, turns out to be an integer in the case of spin manifolds (that is, manifolds whose second Stiefel–Whitney class is null). This result naturally falls in the context of K-theory, but its explanation appeared a real mystery at the time.

In January 1962 Isadore Singer decided to spend (at his own expense) a period of time in Oxford. Two days after his arrival, Atiyah and Singer had this exchange:

> Atiyah: Why is the genus an integer for spin manifolds?
> Singer: What's up, Michael? You know the answer much better than I.
> Atiyah: There's a deeper reason.[8]

Singer has a thorough knowledge of differential geometry and analysis, disciplines in which Atiyah was instead less well-versed. The key to the problem was found in a few months: the answer lay in the Dirac operator. As Atiyah tells it:

---

[7] See N. Hitchin, "Geometria a Oxford: 1960–1990", in *La matematica. Tempi e luoghi*, vol. 1, ed. by C. Bartocci and P. Odifreddi, Einaudi, Torino, 2007, pp. 711–734; much information about the genesis of the index theorem was drawn from this essay.

[8] N. Hitchin, "Geometria a Oxford: 1960–1990", cit., p. 715, quoted from I. M. Singer, "Letter to Michael", in *The Founders of Index Theory. Reminiscences of Atiyah, Bott, Hirzebruch, and Singer*, ed. by S.-T. Yau, International Press, Sommerville 2003, pp. 296–297.

I knew the formula, Hirzebruch's work, I knew what the answer was; what I had to guess was the problem. We had to find out what was this object. We knew from algebraic geometry, what it should look like in algebraic cases. We knew that it had to do with spinors because of Hirzebruch's formula. So the question was, there should be some differential equation which would play the role of the Cauchy–Riemann equations in the spinor case which ought to fit the left-hand side of the equation.[9]

The differential operator "rediscovered" by Atiyah and Singer, whose construction is based on an in-depth study of the geometric properties of spin manifolds, is closely related to the operator, well known to physicists for about 30 years, that appears in Dirac's equations. Atiyah writes:

My knowledge of physics was very slim, despite having attended a course on Quantum Mechanics by Dirac himself, Singer had a better background in the area but in any case we were dealing with Riemannian manifolds and not Minkowski space, so that physics seemed far away. In a sense history was repeating itself because Hodge, in developing his theory of harmonic forms, had been strongly motivated by Maxwell's equations. Singer and I were just going one step further in pursuing the Riemannian version of the Dirac equation. Also, as with Hodge, our starting point was really algebraic geometry.[10]

The result Atiyah and Singer arrived at was admirably simple: on compact spin manifolds, the $\hat{A}$-genus is equal to the index of the Dirac Operator. This not only solves the particular problem, but also defines a unitary framework in which to interpret theorems already known. To clarify this important point, a brief digression is necessary.

A bounded linear operator between Hilbert spaces L: $H_1 \rightarrow H_2$ is said to be a Fredholm operator if its kernel and cokernel (which coincide, we recall, with the kernel of the adjoint operator) both have finite dimensions. The index of the operator is by definition the difference of these dimensions: ind(L) = ker(L) – coker(L). If we consider a continuous family of Fredholm operators, even if the dimensions of the kernels and cokernels vary, their difference remains constant: the index is thus a topological invariant. The Dirac Operator D constructed by Atiyah and Singer for a spin manifold M is the elliptic differential operator between spaces of spinor fields, D: $\Gamma(S^+) \rightarrow \Gamma(S^-)$. If the manifold is compact, D extends to a Fredholm operator between Hilbert spaces, D: $L_2(S^+) \rightarrow L_2(S^-)$; then,

$$\text{ind(D)} = \hat{A}(M).$$

This formula is the prototype of many other analogous formulas: if P is an elliptic operator on a differential manifold M that is compact and oriented, then ind(P) = topological index(P), where the topological index is calculated in terms of appropriate characteristic classes of M (that is, topological data) and of the symbol of P (which is an element in the K-theory of M). The simplest example of this formula is found in the case of a compact Riemann manifold M and operator d+d*, where d is the usual Cartan exterior differential and d* its (formal) adjoint. The

---

[9]M. F. Atiyah, http://www.peoplesarchives.com, cit. (Part 8, 43).

[10]M. F. Atiyah, "Papers on Index Theorem 56–93a", in *Collected Works*, vol. 4, Oxford University Press, New York, 1988, p. 1.

self-adjoint elliptic operator d+d* maps the differential forms of even degree to the differential forms of odd degree: its kernel is the space of the harmonic even forms and its cokernel the space of the harmonic odd forms. Applying Hodge's theorem, we thus find that the index ind(d+d*) is the Euler characteristic of the manifold M; since the topological index d+d* is given by the Euler class of the bundle tangent to M, which is expressed in terms of the curvature of M, the formula ind(d+d*) = topological index (d+d*) is nothing but a re-formulation of the classic Gauss–Bonnet theorem. On manifolds of dimension 4m, the differential forms can be decomposed into two spaces using the Hodge star operator; the same operator d+d* maps one space into another and yields the Hirzebruch signature theorem. In the case of a complex manifold M with a Hermitian metric, one considers the operator $\bar{\partial} + \bar{\partial}^*$, where $\bar{\partial}$ is the Cauchy–Riemann operator (the kernel of $\bar{\partial}$ as an operator on the space of differentiable functions is the space of holomorphic functions). The operator $\bar{\partial} + \bar{\partial}^*$ is elliptic: its index is the holomorphic Euler–Poincaré characteristic of M, and the formula ind($\bar{\partial} + \bar{\partial}^*$) = topological index($\bar{\partial} + \bar{\partial}^*$) exactly reproduces the formula of Hirzebruch–Riemann–Roch.

In the spring of 1962 Atiyah and Singer were therefore able to state the index theorem for spin manifolds. The task of proving it, however, was very difficult. It was just at that time that Stephen Smale, returning from a period spent in Moscow, passed through Oxford, bringing valuable information: in 1959 I. M. Gel'fand had written a fundamental article on the index of elliptic operators,[11] and several Russian mathematicians – for example, M. S. Agranovič and A.S. Dynin – were working on the problem from a very general point of view. As Hitchin observed:

> The advantage [Atiyah and Singer] had over the Russians was that they were concentrating on a particular operator, the Dirac operator, and they knew what the answer should be. They also knew the answer for related operators such as the signature operator and the Dolbeault operators on a complex manifold. The index theorem for each of these cases would give new proofs of the Hirzebruch signature theorem and the Riemann–Roch theorem respectively. Perhaps more importantly, they had seen the problem in the context of K-theory, and that was where the link really lay – the index of an elliptic operator only depends on its highest order term, the principal symbol, and this immediately defines a K-theory class.[12]

Atiyah broadened his own knowledge of analysis, immersing himself in the study of the treatise by Dunford and Schwartz (*Linear Operators*) and other books:

> They were the first books I'd actually tried to read since I was a student. After you've ceased being a student you don't usually read textbooks; you learn what you need to on the hoof.[13]

After the theorem had been announced in the summer at the *Arbeitstagung* in Bonn and at the Stockholm Congress, Atiyah and Singer – calling on the help of friends as well as distinguished mathematicians working in analysis, such as L. Hormänder and L. Nirenberg – achieved the first proof in autumn of 1962, while

---

[11]I. M. Gel'fand, "On elliptic equations", *Russian Math. Surveys*, 15, 3 (1960), pp. 113–123.

[12]N. Hitchin, "Geometria a Oxford: 1960–1990", cit., p. 716.

[13]M. F. Atiyah, http://www.peoplesarchives.com, cit. (Part 8, 45).

spending a period at Harvard. The proof follows one formulated in 1953 by Hirzebruch for his signature theorem, and is therefore based on the theory of cobordism and on (appropriately extended) techniques used in boundary value problems.[14] In spite of the importance of the result, Atiyah was not completely satisfied:

> What was wrong with the first proof, besides being sort of conceptually a bit unattractive ...was that it didn't include some generalisations that we had in mind.[15]

In the coming years Atiyah and Singer would formulate a different proof, based on Grothendieck's proof of the generalised Hirzebruch–Riemann–Roch theorem. In the monumental series of five articles published between 1968 and 1971 in the *Annals of Mathematics* – four in collaboration with Singer, and one with G. B. Segal[16] – Atiyah then went on to formulate various generalizations of the index theorem. These included the equivariant version of the index theorem (in the case of a compact group that acts preserving the elliptic operator) and the version for families of elliptic operators. Together with R. Bott, Atiyah further proved the index theorem for a manifold with boundary and obtained a fixed point formula, which from a certain point of view generalised that of Lefschetz, and from which is obtained as a particular case of Hermann Weyl's celebrated character formula describing the characters of irreducible representations of compact Lie groups in terms of their highest weights.

This impressive crop of results earned Atiyah the Fields Medal in 1966 (shared with P. Cohen, A. Grothendieck and S. Smale[17]). The index theorem is one of the high points of the mathematics of the twentieth century, fundamental not only in its own right, but also because of the multiplicity of its implications and applications, as the justification for awarding the 2004 Abel Prize to Atiyah and Singer makes clear:

> [They were awarded the prize] for their discovery and proof of the index theorem, bringing together topology, geometry and analysis, and their outstanding role in building new bridges between mathematics and theoretical physics The index theorem was proved in the early 1960s and is one of the most important mathematical results of the twentieth century. It has had an enormous impact on the further development of topology, differential geometry and theoretical physics. The theorem also provides us with a glimpse of the

---

[14]F. M. Atiyah and I. M. Singer, "The Index of Elliptic Operators on Compact Manifolds", *Bulletin of the American Mathematical Society* 69 (1963): 422–433. The details of this first proof are given in the volume by R. S. Palais *Seminar on the Atiyah–Singer Index Theorem*, Annals of Mathematical Studies 57, Princeton University Press, Princeton, 1965.

[15]M. F. Atiyah, http://www.peoplesarchives.com, cit. (Part 8, 47).

[16]F. M. Atiyah and I. M. Singer, "The Index of Elliptic Operators" I: *Annals of Mathematics* 87 (1968), pp. 484–530; III: *Annals of Mathematics,* 87 (1968): 546–604; IV: *Annals of Mathematics,* 93 (1971), pp. 119–138; V: *Annals of Mathematics,* 93 (1971), pp. 139–149; F. M. Atiyah and G. B. Segal, "The Index of Elliptic Operators II", *Annals of Mathematics,* 87 (1968), pp. 531–545.

[17]The Fields Medal is awarded to mathematicians who are not yet over 40 years of age: Singer, born in 1924, was therefore ineligible.

beauty of mathematical theory in that it explicitly demonstrates a deep connection between mathematical disciplines that appear to be completely separate.[18]

Each new proof of the index theorem opened unexpected perspectives for research. The works of V. K. Patodi, P. B. Gilkey and Atiyah–Patodi–Singer, going back to the first half of the 1970s, showed that, for a classic elliptic operator (for example, the Dirac operator), the formula for the index can be derived from the study of the asymptotic behaviour of the so-called "heat kernel" associated with the operator. In 1982 Edward Witten, on the basis of physical reasons, discovered a new approach to the problem founded on ideas of symplectic geometry and supersymmetry, which would turn out to be fruitful.[19] The plurality of points of view from which it is possible to consider the index theorem is a further proof of the conceptual depth of this result. In fact, Atiyah says:

> Any good theorem should have several proofs, the more the better. For two reasons: usually, different proofs have different strengths and weaknesses, and they generalize in different directions – they are not just repetitions of each other. And that is certainly the case with the proofs that we came up with. There are different reasons for the proofs, they have different histories and backgrounds. Some of them are good for this application, some are good for that application. They all shed light on the area. If you cannot look at a problem from different directions, it is probably not very interesting; the more perspectives, the better![20]

Geometry and Physics

In 1969 Atiyah left Great Britain, accepting a position as professor at the Institute for Advanced Studies in Princeton, where he would stay 3 years. He returned to Oxford in 1973 as Royal Society Research Professor and Fellow of St. Catherine's College. In 1990 Atiyah – who in 1983 had been knighted and granted the title of Sir – transferred to Cambridge, becoming Master of Trinity College and director of the newly-founded Isaac Newton Institute for Mathematical Sciences. From 1990 to 1995 he was president of the Royal Society.

Starting in 1977, Atiyah's research interests gradually moved towards gauge theory and, more generally, towards the interaction between geometry and physics. He was first urged to consider problems of mathematical physics by Roger Penrose, who had been his fellow Ph.D. student at Cambridge (in those years both did research in algebraic geometry and for a certain period of time both had Hodge

---

[18]As quoted in N. Hitchin, "Geometria a Oxford: 1960–1990", cit., p. 717.

[19]Witten's ideas would be developed by, among others, L. Alvarez-Gaumé, E. Getzler, N. Berline, M. Vergne and J. P. Bismut. The expert reader can consult N. Berline, E. Getzler and M. Vergne, *Heat Kernels and Dirac Operators*, Springer-Verlag, Berlin, 1992.

[20]M. Raussen and C. Skau, "Interview with Michael Atiyah and Isadore Singer", *Notices of the American Mathematical Society*, 52 (2005), pp. 223–231.

as supervisor) and in 1973 became Rouse Ball Professor of Mathematics at Oxford. The two had long discussions about twistor theory – a powerful instrument for studying some equations in mathematical physics invented by Penrose and, at the time, thought incomprehensible by many (Freeman Dyson, for example, had said, "Twistors are a mystery"). Atiyah had no difficulty understanding twistor geometry, since this was based on Klein's classic correspondence for straight lines in the complex projective space $\mathbb{P}^3$, and he explained to Penrose how to use techniques of sheaf cohomology for computing some complicated contour integrals.[21] Together with Richard Ward, one of Penrose's Ph.D. students, he sought to the interpret self-dual Yang–Mills equations[22] in terms of twistors: in a joint article of 1977, Atiyah and Ward defined a correspondence – today called Atiyah–Ward correspondence – between instantons on the four-dimensional sphere and certain holomorphic bundles on $\mathbf{P}^3$.

In that same year Singer was also in Oxford on sabbatical, and helped turn Atiyah's interests in the direction of Yang–Mills equations. The fundamental article "Self-duality in four-dimensional Riemannian geometry"[23] by Atiyah, Nigel Hitchin and Singer, besides developing in detail twistor theory in the context of Riemannian geometry, introduced the instruments essential for studying the moduli space of Yang–Mills instantons (the dimension of this space is computed by applying an appropriate version of the index theorem). Still in 1977, Atiyah and Hitchin tackled the problem of describing all of the instantons on the S^4 sphere in terms of linear data, using some objects (monads) introduced in algebraic geometry by Barth and Horrocks to study holomorphic bundles on complex projective spaces. As Hitchin remembers it:

> The pieces of the jigsaw were finally assembled by Atiyah and the writer before going off to have lunch at St. Catherine's College on November 22 1977. On our return to the Mathematical Institute we found a letter from Y. Manin giving essentially the same construction with V. G. Drinfel'd. A joint paper was published and the method became known as the ADHM construction of instantons.[24]

---

[21]See M. F. Atiyah, "Papers on gauge theories", in *Collected Works*, vol. 5, Oxford University Press, New York, 1988, p. 1.

[22]The Yang–Mills theory is a gauge theory, whose Lagrangian is written in terms of the curvature of a principle bundle of structural group G; the Euler–Lagrange equations of the corresponding action functional are the Yang–Mills equations. If the base manifold is the Minkowski space and G=(U)1, then the Yang–Mills equations are simply Maxwell's equations. In the case where G=SU(2)xU(1), the Yang–Mills equations are classic (that is, not quantum) field equations of Glashow–Weinberg–Salam's electroweak theory. In mathematical physics, the Yang–Mills theory is studied on a generic Riemannian manifold of dimension 4 (therefore not in Lorentzian signature). The Yang–Mills equations are not linear; the self-dual or anti-self-dual Yang–Mills equations are a linearization of them, whose solutions are called instantons or anti-instantons (which correspond to the absolute minima of the action functional).

[23]*Proceedings of the Royal Society of London*, series A, 362 (1978), pp. 425–461.

[24]N. Hitchin, "Geometria a Oxford: 1960–1990", cit., p. 723. On the ADHM construction, see M. F. Atiyah, *Geometry of Yang–Mills fields*, Lezioni Fermiane, Scuola Normale Superiore, Pisa, 1979.

The first half of the 1980s was a period of an extraordinary flowering in geometry at Oxford. Atiyah headed a group that included numerous mathematicians – including, among others, Segal, Hitchin, George Wilson, and later, Simon Salamon and Dan Quillen – and a large company of Ph.D. students, many of whom were destined to become first-rank scientists, above all Simon Donaldson (first a student of Hitchin, and later of Atiyah). The group met each Monday afternoon at 3 o'clock for the seminar on "geometry and analysis" organized by Atiyah: the lectures of famous mathematicians from the world over alternated with those of Atiyah himself, "which were invariably virtuoso performances".[25] The research topics – above all those carried out by the Ph.D. students – were predominantly determined by the results obtained by Atiyah in the area of geometry of gauge theories: besides the articles already mentioned, at least two others were of fundamental importance. Atiyah's influential paper with Bott concerning Yang–Mills equations over a Riemann surface[26] – "extraordinarily wide-ranging and many-faceted"[27] – introduces a large number of innovative ideas (for example, the interpretation of the moduli space as a symplectic quotient through the construction of a "moment map" in an infinite-dimensional setting). In Atiyah's article in collaboration with John D. Jones, which studies the topology of the moduli spaces of instantons, is instead stated the famous conjecture that bears the name of both authors.[28] In this extremely stimulating environment – in which "technical specialisation, as an algebraic geometer, differential geometer, topologist or whatever, was not particularly encouraged" and "the great thing was to explore the interaction of these different areas"[29] – Simon Donaldson obtained, using the ideas and methods of the Yang–Mills theory, the spectacular results on the geometry of the four-dimensional differentiable manifolds which earned him the Fields Medal in 1986.

What was gradually outlined, at least in regard to its basic characteristics, was the scheme that would reveal deep and unexpected connections between geometry and physics. The interaction between the two disciplines brings into play, on one side, the quantum aspects of field theories, and on the other, the global topological properties of geometric objects. An example of the effectiveness of this perspective

---

[25] S. K. Donaldson, "Geometry in Oxford c. 1980–1985", *Asian Journal of Mathematics* 3 (1999), pp. xliii–xlviii. We have also drawn from this article information regarding the principal lines of research in geometry at Oxford in those years.

[26] M. F. Atiyah and R. Bott, "The Yang–Mills equations over Riemann surfaces", *Philosophical Transactions of the Royal Society of London*, series A, 308 (1982), pp. 523–615.

[27] S. K. Donaldson, "Geometry in Oxford c. 1980–1985", cit., p. xliv.

[28] M. F. Atiyah and J. D. S. Jones, "Topological aspects of Yang–Mills theory", *Communications in Mathematical Physics* 61 (1978) 97–118. The Atiyah–Jones conjecture asserts that, given a principal SU(2) bundle over a 4-manifold X, the inclusion of the moduli space of framed instantons of charge k into the space of all gauge equivalence classes of connections of the same charge induces an isomorphism in homotopy and homology through a range that grows with k. The conjecture was proven for S^4 (by Boyer, Hurtubise, Milgram and Mann in 1993) and for various other classes of 4-manifolds, but not in its generality.

[29] S. K. Donaldson, "Geometry in Oxford c. 1980–1985", cit., p. xliv.

is given by the proof of the Morse inequalities obtained by Witten in 1982, making use of ideas drawn from supersymmetric field theories.

In effect, quantum field theory provides new and powerful instruments for investigating the geometry of manifolds in dimensions 2, 3 and 4. The differentiable invariants of 4-manifolds defined by Donaldson in 1988, and Andreas Floer's theory for manifolds of three dimensions, are interpreted by Witten in a unified framework: the Donaldson–Floer theory can be described as a topological quantum field theory (TQFT) in 3+1 dimensions. Another case that is just as important is given by the invariants of knots discovered by Vaughan Jones in 1987.[30] "These are related to physics in various ways but the most fundamental is due to Witten who [shows] that the Jones invariants have a natural interpretation in terms of a topological quantum field theory in 2+1 dimensions".[31] Atiyah played a prominent role in these developments: he not only obtained important results (for example, he defined an axiomatics for TQFTs that takes up and elaborates on Graeme Segal's work), but he also spared no effort to spread the new ideas – especially those of Witten – contributing to their acceptance by the mathematical community, initially inclined to consider them hard and unorthodox.

Conclusion

The interaction between mathematics and physics has been one of the principal driving forces of mathematics in the past 30 years: from gauge theories to string theories, from supersymmetry to the theory of integrable systems. Atiyah poured all of his scientific authority, his inexhaustible energy and his contagious enthusiasm into promoting this interaction. His enormous influence on the international mathematical community goes beyond his scientific work: he has created about him – tirelessly discussing with mathematicians and physicists, persisting in his search for "the deepest reasons" that underlie theorems – a movement of ideas that has strongly directed the research of the last three decades; he has had dozens of Ph. D. students, some of whom have become top level mathematicians; instantly seeing their originality and significance, he has increased the prestige of the results of Andreas Floer, Vaughan Jones, Edward Witten and several others; and he has disseminated his own conception of mathematics in hundreds of lectures.

In his speech on the conferment of the Feltrinelli Prize of the Accademia dei Lincei in 1981, Atiyah thus summed up his own course of research and his personal vision of mathematics as a "social activity":

---

[30]Vaughan Jones and Witten were awarded the Fields Medal in 1990. Andreas Floer, to whom are owing important results in topology and in symplectic geometry (including the definition of the so-called Floer homology and the proof, in a special case, of the Arnol'd conjecture on fixed points of a symplectomorphism), tragically ended his own life in 1991 at 45.

[31]M. F. Atiyah, *The geometry and physics of knots*, Lezioni Lincee, Cambridge University Press, Cambridge, 1990, p. 2.

[...] my mathematical interests have gradually shifted from field to field, starting with algebraic geometry and ending up with theoretical physics. On the other hand the change was never a deliberate or discontinuous one. It was simply that the problems I studied naturally led me in new directions, frequently into quite foreign territory. Moreover the link between the different areas was an organic one, so that I could not discard the old ideas and techniques when moving into a new field – they came with me. [...] most of my work has been carried out in close and extended collaboration with mathematical colleagues. I find this the most congenial and stimulating way of carrying on research. The hard abstruseness of mathematics is enlivened and mollified by human contact. In addition the very diversity of the fields in which I have engaged has made it essential to work with others. I have indeed been fortunate in having had so many excellent mathematicians as my friends and collaborators.[32]

---

[32]M. F. Atiyah, "Speech on Conferment of Feltrinelli Prize", in *Collected Works*, vol. 1, Oxford University Press, New York, 1988, pp. 315–316.

Vladimir Igorevich Arnold

Universal Mathematician

Marco Pedroni

The list of fields of mathematics in which Vladimir Igorevich Arnold (born in 1937 in Odessa, passed away in 2010 in Paris) made fundamental contributions is very long. If, without any attempt at being exhaustive, we limit ourselves to geometry and mathematical physics, we have algebraic geometry (real and complex), symplectic topology and the geometry of contact varieties on the one hand, and on the other hydrodynamics, classical mechanics, celestial mechanics, integrable systems and the theory of dynamical systems. His name is tied to many key concepts in twentieth-century mathematics and mechanics, such as the Kolmogorov–Arnold––Moser (KAM) theory, Arnold diffusion, Arnold-stability (or A-stability, in hydrodynamics), and the characteristic classes of Arnold–Maslov, to name only a few. This belies what he himself calls "Arnold's law", according to which only a very small number of discoveries are attributed to the right person.

The contribution which made him famous throughout the world at only 20-years old was the solution of Hilbert's 13th problem, which regards whether it is possible or not to solve algebraic seventh-degree equations using functions with two arguments. To be more precise, the question was the following: can the real function $z(a,b,c)$ defined by the equation $z^7 + az^3 + bz^2 + cz + 1 = 0$ be represented as a composition of continuous functions of two variables? In his doctoral thesis, written with Kolmogorov as his advisor, Arnold gave a positive answer to this question, proving that every continuous function with three variables can be constructed starting with functions of only two variables.

Following this he dedicated himself to dynamical systems, making a determinant contribution to the creation of what would come to be famous as KAM theory. In this case the starting point was the study of integrable systems, which are Hamiltonian systems whose behaviour is very regular: given suitable hypotheses, the motion of this kind of system is quasi-periodic, that is, it is composed of uniform rotations. From the point of view of geometry, the system's phase space ($2n$-dimensional, where n is the number of degrees of freedom) turns out to be the union of n-dimensional tori – called invariant tori – and the motion of the system takes place uniformly on these tori. Many systems of practical interest, above all that of the solar system, can be seen as mere perturbations of an integral system.

C. Bartocci et al. (eds.), *Mathematical Lives*,
DOI 10.1007/978-3-642-13606-1_29, © Springer-Verlag Berlin Heidelberg 2011

This is why Poincarè considered the study of the behaviour of a perturbed integrable system to be the fundamental problem in dynamics. In 1954 Kolmogorov announced a surprising result: if the perturbation is small and the initial integrable system is not degenerate, then the majority of the invariant tori remain as such and the motion on them is quasi-periodic. Arnold not only provided a complete and rigorous exposition of the proof of Kolmogorov's theorem; he also generalised it to a broad class of degenerate systems and presented numerous applications to classical problems of dynamics.

Vladimir Igorevich Arnold

Another field in which Arnold's genius can be seen clearly is that of hydrodynamics. Taking advantage of the analogy with the inertial motions of a rigid body with a fixed point, he proved that Euler's equations (which describe the motion of an ideal fluid) can be interpreted as the geodesic equations – with respect to a metric defined by kinetic energy – of the group of diffeomorphisms preserved by the volumes. This provides an explanation of why the motions of atmospheric masses are instable and thus why it is so difficult to make reliable long-range weather forecasts: the curvatures of these groups are negative, and thus two geodesics that are initially close to each other rapidly move away from each other. By using methods of topology, Arnold then classified the stationary motions of a fluid (in the plane and in the space) and found the sufficient conditions for their stability.

Arnold also shed light on the symplectic nature of a theory conjectured by Poincaré and proved by Birkhoff. This says that a diffeomorphism of a circular corona having the property of preserving the areas and rotating the two borders in opposite directions, has at least two fixed points. Arnold hypothesised that this is a particular case of the fact that a symplectomorphism homologous to the identity has a number of fixed points that is greater than or equal to the sum of Betti numbers, a topological invariant of the manifold on which the symplectormorphism acts. This result was then proved and marked the birth of symplectic topology, as well as becoming the starting point leading to the discovery of Floer homologies and quantum cohomologies. We conclude this brief recap of his results with mention of the fact that Arnold, driving by problems of quantum optics, classified the (symplectic) singularities of functions with n real variables, showing that these are related to the Dynkin diagrams that also appear in the classification of simple Lie algebras.

All across the globe, many university students of mathematics and physics have had the good fortune of using Arnold's books, especially *Geometrical Methods in the Theory of Ordinary Differential Equations* and *Mathematical Methods of Classical Mechanics*. Arnold favoured a presentation in which the ideas, examples, motivations (often taken from physics) and geometric intuitions play a leading role, more than that of a rigorous treatise that is cold and arid. His article "On teaching mathematics" begins with the words, "Mathematics is a part of physics. Physics is an experimental science, a part of natural science. Mathematics is the part of physics where experiments are cheap". This is why he is perennially opposed to the Bourbakian concept of mathematics teaching and strenuously defends the point of view of nineteenth-century mathematics (it appears to be thanks to Arnold that the works of Goursat and Hermite, which were supposed to be eliminated from the library of one French university, were preserved).

Arnold's research has been honoured with many international prizes (including the Lenin Prize in 1965, with Andrey Kolmogorov, the Crafoord Prize in 1982 with Louis Nirenberg, and the Wolf Prize in Mathematics in 2001). He was a member of numerous scientific academies (including the Accademia dei Lincei, since 1988). He was vice-president of the International Mathematical Union from 1995 to 1998 and has received honorary degrees from universities the world over, including one from the University of Bologna in 1991. He has had an enormous number of students, many of which have gone on to become first-class mathematicians and have contributed to the spread of his ideas and his unifying approach to mathematics (and physics), his problems, and his teachings.

Enrico Bombieri

The Talent for Mathematics

Enrico Bombieri was born in Milan in 1940. His precocious talent for mathematics was supported by his family, and while still a boy he came into contact with some eminent mathematical scholars. One of these was Giovanni Ricci, who worked in analysis and number theory; his influence would be a determining factor in Bombieri's development. During these years Bombieri laid the foundations for his vast and in-depth knowledge of classical mathematics, which would be one of his distinguishing traits. He published his first work, concerning the solution to a Diophantine equation, in 1957, while still in high school; when he enrolled in mathematics at the University of Milan he already had the maturity of a professional mathematician. During his years at university, his name began to circulate in international mathematics circles; before he graduated, under Ricci's supervision, he produced numerous results in various areas in number theory and complex analysis, some of which were of significant importance and made quite an impact on the mathematical community. One example dealt with the growth of the remainder term in the elementary proof of the prime number theorem. During this period Bombieri visited Trinity College in Cambridge to study with Harold Davenport, a distinguished mathematician and excellent supervisor, another key figure in his scientific formation.

In 1965 Bombieri published a fundamental result on the distribution of primes in arithmetic progressions, which can be used in applications in place of Riemann's hypothesis. The work was based on a new development of the large sieve introduced by Yuri V. Linnik in 1941, and signalled a turning point in analytical number theory. In that same year, Bombieri was given the chair in mathematical analysis, and after a brief stint at the University of Cagliari, was called to the University of Pisa. This was a very fruitful period for his scientific work: in addition to his profound contributions in number theory, there were others regarding univalent functions and Bieberbach's conjecture, partial differential equations, minimal surfaces (in particular, the solution to Bernstein's problem for higher dimensions), the algebraic values of meromorphic functions of several variables, and the

C. Bartocci et al. (eds.), *Mathematical Lives*,
DOI 10.1007/978-3-642-13606-1_30, © Springer-Verlag Berlin Heidelberg 2011

classification of algebraic surfaces. He also collaborated with a number of leading mathematics such as Harold Davenport, Peter Swinnerton-Dyer, Ennio De Giorgi, Aldo Andreotti and David Mumford. In 1973 he was called to the Scuola Normale Superiore in Pisa, and in 1974, first and so far unique Italian mathematician, he was awarded the Fields Medal. In 1980 he transferred to the United States as a permanent member of Princeton's Institute for Advanced Study, where he works today. That same year he won the Balzan Prize. In 2000 he was given the honoris causa doctor's degree from the University of Pisa.

Enrico Bombieri

Bombieri still carries on his incessant research activities. On the one hand, he has broadened his spectrum to include important contributions in Diophantine approximation, finite group theory, and Diophantine geometry; on the other, he goes back, especially in collaboration with Henryk Iwaniec, to classic themes in analytical number theory such as the distribution of primes, the Riemann zeta function, and the theory of L-functions. In 1996 he was elected a member of the National Academy of Sciences. In 2000, the centennial of Hilbert's famous 23 problems, presented at the International Congress of Mathematicians in Paris in 1900, the Clay Mathematics Institute established a prize of a million dollars a piece for solutions to the seven "Millennium Prize Problems". Bombieri was named to make the official presentation of Riemann's hypothesis, which is probably the most famous and important open problem in contemporary pure mathematics.[1] In 2006 he won the International Pythagoras Prize for mathematical research, which was instituted by the Municipality of Crotone only in 2004 but whose pedigree is impeccable (the prize was awarded to Andrew Wiles in 2004, Edward Witten in 2005).

Bombieri's mathematical talent has been aptly described in the official presentations of some of the honours he has received. On the occasions of the presentation to Bombieri of the Fields Medal, Kamaravolu Chandrasekharan wrote,

> I hope I have said enough to show that Bombieri's *versatility* and *strength* have combined to create many original patterns of ideas which are both rich and inspiring. ... To him mathematics is a private garden; may it bring forth many new blooms.[2]

The official text of his nomination to the National Academy of Sciences says,

> Bombieri is one of the world's most versatile and distinguished mathematicians. He has significantly influenced number theory, algebraic geometry, partial differential equations, several complex variables, and the theory of finite groups. His remarkable technical strength is complemented by an unerring instinct for the crucial problems in key areas of mathematics.[3]

To this it must be added that Bombieri also had considerable talent as a presenter; his lectures and writings are always fascinating and offer a knowing mixture of clarity and synthesis. During an interview he said:

> To become a mathematician was for me like to follow a calling full of satisfactions, that never tires you. It is always a great joy to achieve the understanding of a new theory and imagine the beauty and usefulness it can give.

Finally, Bombieri has many other interests besides mathematics, from poetry to cuisine, but above all painting, which he considers his second profession.

In his introduction to the Italian edition of André Weil's *Theory of Numbers* (*Teoria dei numeri*, Einaudi, Torino, 1993) Enrico Bombieri wrote:

> Mathematics is an art that contains its own justification and foundation, in the same way that Michelangelo's sculptures live inside the stone until they are liberated by the chisel.

This sense of beauty, which is connected to an extraordinary virtuosity, led Bombieri to the solution of many of the key problems in today's mathematics.

---

[1] See *The Millennium Prize Problems*, J. Carlson, A. Jaffe, A. Wiles, eds., Clay Mathematics Institute and the American Mathematical Society.

[2] K. Chandrasekharan, The work of Enrico Bombieri, *Proceedings of the International Congress of Mathematicians*, Vancouver 1974 1 (Montreal, Que., 1975), 9–10.

[3] Cf. The National Academy of Sciences website: http://www.nasonline.org.

Martin Gardner

The Mathematical Jester

Ennio Peres

Martin Gardner, the world's most authoritative and prolific author of works on mathematical recreation of all time, was born on 21 October 1914 in Tulsa, Oklahoma and died on 22 May 2010. From 1956 to 1981 he was editor of a column in *Scientific American* dedicated to mathematical puzzles and games which soon had a fans in every corner of the globe. He has also published hundreds of articles in various magazines and written more than 70 books dealing with topics ranging from science to philosophy, from mathematics to literature.

One of Gardner's main characteristics is that, with a gamer's lightness of touch, he is always able to enter into even the most complex branches of mathematics, starting from points that are curious and intriguing. Contrary to what you might think, however, his formal training was not in the sciences. His only degree is in philosophy, which he received from the University of Chicago in 1936. He is completely self-taught in mathematics. His desire to know grew out of his lifelong passion for magic tricks, which started when he was just a boy, and an innate curiosity in transcendent themes. His first book, *Fads and Fallacies in the Name of Science*, published in 1952, examines and dismantles more than 50 kinds of pseudoscientific beliefs concerning the paranormal.

Also in 1952, Gardner began to collaborate with the *Humpty Dumpty's Magazine*, a magazine for children, for which, in addition to writing imaginative stories, he also devised original games with paper, such as that shown below.

Figure 1 shows 11 rabbits. If, however, the two rectangles marked A and B switch places, one of the rabbits mysteriously disappears, transforming into an egg, as shown in Fig. 2.

His experience with Humpty Dumpty, which lasted 8 years, honed his skills in writing fanciful narratives, and even more importantly, it accustomed him to writing in a style that was clear, simple and direct.

A few years later, in 1956, just before he began collaborating with *Scientific American*, Gardner published *Mathematics, Magic and Mystery*, an original work that gathers and classifies the most interesting magic tricks on the basis of a mathematical way of thinking, constituting a kind of manifesto of "Mathemagic", a term coined in 1951 by the magician Royal V. Heath.

C. Bartocci et al. (eds.), *Mathematical Lives*,
DOI 10.1007/978-3-642-13606-1_31, © Springer-Verlag Berlin Heidelberg 2011

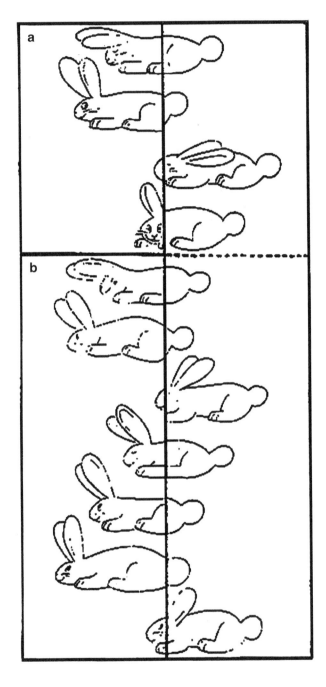

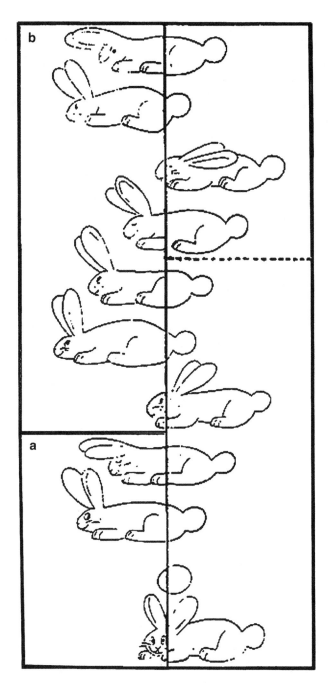

One of the most fascinating problems popularised by Gardner in the course of his long career is the following:

> Jones, a gambler, puts three cards face down on the table. One of the cards is an ace; the two others are face cards. You place a finger on one of the cards, betting that it is an ace. Obviously, the probability that it actually is an ace is equal to 1 out of 3. Jones now secretly peeks at the three cards. Given that there is only one ace, at least one of the cards you didn't choose has to be a face card. Jones turns it over and shows it to you. At this point, what is the probability that you've got your finger on the ace?

Solution

Many think that at this point the probability has risen from 1 out of 3 to 1 out of 2. After all, there are only two cards face down, and one has to be the ace. But actually, the probability remains 1 out of 3. The probability that you have not picked the ace remains 2 out of 3, even if Jones seems to have eliminated part of the uncertainty by showing you that one of the two remaining cards was not the ace. However, the probability that the other of the two cards remaining is the ace is still 2 out of 3, because the choice was made before. If Jones were now to give you the chance to change your bet, you'd do well to accept (as long as you don't have any aces up your sleeve!).

Martin Gardner presented this problem for the first time in *Scientific American* in October 1959, under a different form (instead of three cards, there were three prisoners, one of who had received a pardon). In 1990 Marilyn vos Savant, author of a popular column in *Parade* magazine, proposed yet another version involving three doors, behind which were hidden a car and two goats. She gave the correct answer, but she received thousands of letters from irate readers (including many sent by mathematics teachers!) accusing her of not knowing anything about probability theory. The case ended up on the front page of the *New York Times*, and in a short time the problem became famous all over the world. It finally came to be considered the most intriguing probabilistic paradox of the second millennium.

Le Corbusier's Door of Miracles

Mathematics is the majestic structure conceived by man to grant him comprehension of the universe. It holds both the absolute and the infinite, the understandable and the forever elusive. It has walls before which one may pace up and down without result; sometimes there is a door: one opens it – enters – one is in another realm, the realm of the gods, the room which holds the key to the great systems. These door are the doors of the miracles. Having gone through one, man is no longer the operative force, but rather it is his contact with the universe. In front of him unfolds and spreads out the fabulous fabric of numbers without end. He is in the country of numbers. He may be a modest man and yet have entered just the same. Let him remain, entranced by so much dazzling, all-pervading light.

The shock of this light is difficult to bear. The young, who bring us the support of their enthusiasm and that unawareness of responsibility which is the strength and weakness of their age, envelop us – if we do not resist it – in the mists of their uncertainties. In the matter with which we are dealing here, it is necessary to stand firm and know what we are seeking: a precision instrument to be used for the choice of measures. Once the compasses are in our hands, ploughing the furrow of numbers, the roads begin to ramify, spread out in all directions, flourish and multiply ... and carry us away, veering from the aim we have set ourselves: the numbers are at play! ... Architecture is not a synchronic phenomenon but a successive one, made up of pictures adding themselves one to the other, following each other in time and space, like music.

"Music is a secret mathematical exercise, and he who engages in it is unaware that he is manipulating numbers". (Leibniz.) ... music rules all things, it dominates; or, more precisely, harmony does that. Harmony reigning over all things, regulating all the things of our lives, is the spontaneous, indefatigable and tenacious quest of man animated by a single force: the sense of the divine, and pursuing one aim: to make a paradise on earth. In Oriental civilizations, "paradise" meant a garden: a garden beneath the rays of the sun or in the shade, shimmering with the most beautiful of flowers, glowing with a wealth of green. Man can only think and act in terms of man (the measures which serve his body) and integrate himself in the universe (a rhythm or rhythms which are the breathing of the world).

C. Bartocci et al. (eds.), *Mathematical Lives*,
DOI 10.1007/978-3-642-13606-1_32, © Springer-Verlag Berlin Heidelberg 2011

Extracted from: Le Corbusier, *The Modulor: A Harmonious Measure to the Human Scale Universally applicable to Architecture and Mechanics*, translated from the French into English by Peter de Francia and Anna Bostock, Birkhäuser Publishers, Basel, 2000, pp. 71–74. Reproduced by kind permission of Birkhäuser Publishers.

F. William Lawvere

The Unity of Mathematics

Renato Betti

The first lesson learned when approaching Bill Lawvere's work and intense activity is that in mathematics there are no facts that are specific, unique and meaningful for only a few special cases. The traditional dividing lines between geometry, analysis, mechanics, and so forth are themselves artificial, and while their conventional limits are easily (for him) crossed, their interconnections should always be borne in mind.

The search for unity, of a conceptual context that would render the fundamental notions clear and explicit – of mathematics, but of physics as well, especially continuum mechanics – was a distinguishing characteristic of his scientific work from the very beginning, and is still today a strong component of it.

This tendency grows out of careful observation of the facts – not only those of mathematics – and their organisation according to a precise "guideline" for "understanding, learning, and developing mathematics", as he put it. A conceptual guide whose presence is easily intuited, and whose lines are often made clearly explicit by Lawvere in his work and in his discussions.

What are the foundations of mathematics? Right away Lawvere's thinking took an original course with respect to the traditional logical conception which, according to him, tends in some way to obscure the aspiration to find a conceptual guide for our actions:

> Foundations will mean here the study of what is universal in mathematics. Thus Foundations in this sense cannot be identified with any "starting-point" or "justification" for mathematics, though partial results in these directions may be among its fruits. But among the other fruits of Foundations so defined would presumably be guide-lines for passing from one branch of mathematics to another and for gauging to some extent which directions of research are likely to be relevant.[1]

---

[1]F. William Lawvere, "Adjointness in Foundations", *Dialectica*, no. 23 (1969), pp. 281–296.

C. Bartocci et al. (eds.), *Mathematical Lives,*
DOI 10.1007/978-3-642-13606-1_33, © Springer-Verlag Berlin Heidelberg 2011

F. William Lawvere

Lawvere's conceptual guide is the rejection of the ideological perspective according to which theory is more fundamental than practice. Declaring his own debt to Hegel's philosophy and Engel's considerations, Lawvere calls this guide *the logic of mathematics*, distinct from and comprising *mathematical logic*: it is how our thinking develops, like the science of the shape of space and quantitative relations, as need arises. Mathematical work is thus linked to the investigation of the general laws of thinking, applied to the study of particular subjects.

In this vast program, the important role of mathematics and its logic as unifiers are acknowledged in category theory, a new subject, created in 1945 by S. Eilenberg and S. Mac Lane to solve concrete problems in analysis and algebraic topology – to unify and simplify many phenomena that arose in the 1930s – and later developed above all by A. Grothendieck and his school in the 1950s and 1960s, with a view to providing the foundations of contemporary algebraic geometry.

F. William Lawvere was born in Indiana in 1937. His training developed along lines that were in some way "exemplary". He began in experimental physics, but went from that to theoretical physics because he was looking for a guide to direct his practice. The next step was deciding to work actively with mathematics: better, mathematics connected to physics – more specifically, to rational mechanics – which at that time at the University of Indiana was taught by Clifford Truesdell. He finally arrived to the discovery of category theory, which had not yet been put to the test but which promised to explain mechanics, analysis, geometry, and even how these were related. This was a subject that would render explicit the philosophical approach which it, like all subjects, contained: at the heart of the investigation lay the relationship between objects.

Lawvere had completed his mathematical training in 1963 at Columbia University in New York City under the guidance of a mathematician of renown, Samuel Eilenberg. His doctoral thesis on algebraic theories[2] soon became a point of reference and is by now considered a classic; its fortieth anniversary was celebrated with a conference which included the participation of the major specialists in category theory, logic, foundations of mathematics, as well as experts in the philosophy of science, etc.[3]

In Lawvere's thesis, universal algebra was freed from its artificial dependence on presentations with respect to particular operations, as had been usual with the subject up to that time. A theory is something more objective than one of its possible presentations.

The central idea is that an algebraic (or equational) theory is a category with certain properties (finite products) and that it is useful to describe the models by means of the functors in the category of sets that have those properties. Thus the classic algebraic constructions are expressed in universal terms by means of the properties of adjoint functors.

But this is not just an ad hoc description within a specific area: specificity is not part of Lawvere's perspective. As usually occurs in his work, he opens the way for a unifying comprehension of other areas, of apparently different points of view, of paths that are entirely new and original. These too – as Lawvere has said quite clearly – are fruits of the search for what is fundamental in mathematics.

In the 1960s it was discovered that there are theories in other categories as well that are suited to study in these universal terms: for example, that of first-order (or elementary) theory, based on Lawvere's observation that the existential quantification and universal quantification can be described with adjoint functors respectively to the left and to the right of the substitution operator. Further, it makes sense to consider the models outside of the category of sets as well, and this vast unification leads to ever greater comprehension: thus is born *categorical logic*, in which the main logical theories are carried over to the context of mathematics where they are subject to manipulation and transformation according to the rules of algebra.[4] The entire apparatus of logic – from syntax as well as from semantics – can be conceived of and restated in categorical terms.

This also makes it possible to interpret theories in worlds other than those of sets, developed in parallel because of the need to understand and formalise the intuitive notion of a "continuously variable set", to use the brilliant metaphor coined by

---

[2]The essential points of this thesis were published in the *Proceedings of the National Academy of Sciences* (no. 50, 1963, pp. 869–872) with the title "Functorial Semantics of Algebraic Theories".

[3]"Ramifications in Categories", Florence, 18–22 November 2003.

[4]See for example, "Some Algebraic Problems in the Context of Functorial Semantics of Algebraic Theories", *Lecture Notes* no. 61, 1968, pp. 41–61, Springer-Verlag.

Lawvere himself. This led to the birth of *elementary topoi* at the end of the 1960s, fruit of Lawvere's collaboration with the topologist Myles Tierney:

> ... experience with sheaves, permutation representations, algebraic spaces, etc., shows that a "set theory" for geometry should apply not only to abstract sets divorced from time, space, ring of definition, etc., but also to more general sets which do in fact develop along such parameters.[5]

The inspiration for this came from the ideas and work that Grothendieck and his school were producing in those same years in the context of algebraic geometry, with important results regarding complex geometry and sheaf theory: today a "Grothendieck topos", a category of sheaves for a Grothendieck topology, is a new area which provides a unifying formulation for many situations that occur in algebraic geometry.

However, Lawvere's work soon morphed into the search for what can be defined in elementary terms, independent of any set theory, resulting in an unexpected unification of some aspects of algebraic geometry and corresponding aspects of geometric logic[6]: the idea of *elementary topos*, like that of Grothendieck's topos more generally, reflects precisely the possibility of freeing itself from dependence on sets. This can be traced back to earlier works of Lawvere, for example in an "elementary" characterisation of the category of sets of 1964, and to an analogous description of the category of categories.[7]

This formal notion of "variable set", besides providing an elementary axiomatic of the category of sets, also achieves a conceptual interaction between, and often a genuine unification of, important theories developed independently of one another in those years: Grothendieck's topos, Robinson's non-standard analysis, Kripke semantics (which would later come to be known as Kripke–Joyal semantics, or topos semantics). Cohen's independence theorems would be proven again in set theory, the theory of metric space would be united to the logic of higher-order predicates, the project of rediscovering and rewriting a large part of mathematics in the context of variable sets would begin.

Lawvere also refined and refounded (or founded ex novo) other significant fields in mathematics as well. Here are just some examples: in 1967 he showed that some categories with a particular "infinitesimal" object define contexts in which it is possible to study the models of differential geometry and continuum mechanics in a flexible way. This appeared to be an inconsequential observation, and so it remained for a number of years, until its importance was finally understood (in the 1980s). The idea was developed until it finally gave rise to the important area of research that is known today as "synthetic differential geometry", with applications

---

[5]F. William Lawvere, "Quantifiers and Sheaves", *Actes, Congrès international des mathematiciens,* Nice 1970, Gauthier-Villars (1971), vol. 1, pp. 329–334.

[6]F. William Lawvere, "Continuously Variable Sets: Algebraic Geometry=Geometric Logic", *Proceedings of the Logic Colloquium* (Bristol 1973), North Holland, 1975, pp. 135–157.

[7]"The Category of Categories as a Foundation for Mathematics", La Jolla Conference on Categorical Algebra (1966), pp. 1–20.

in the calculus of variations, differential equations, and the calculation of singularities in the applications to varieties. There's more: some of the concepts that are fundamental in mathematics are defined in formal terms, like *quantity categories* and *space categories*, which are studied by means of *intensive quantities* and *extensive quantities*. Visualising numerous constructions in terms of "unity or identity of opposites", a notion expressed by means of adjunct functors, makes it possible to see that not only is the development of mathematics not independent of a general way of thinking and behaving, but in fact cannot be independent of it. What was important was understanding the particular role played by special objects in special categories: the *object of truth values* in elementary topoi, which make it possible to define the relations and the operations that are partially defined; the *object of natural numbers* in the study of the operations of recursion and induction, whose category characterisation would become known by the name *Peano–Lawvere axiomatisation*; the *differential object* in the category of smooth spaces, and so on.

The search for unity is an important strategy for complex systems, to be pursued for not only in the development of applications, but also in the study of mathematics.

In this sense, for Lawvere, the study of the foundations of mathematics, is intimately connected to problems of mathematics education and training. The unification of mathematics is in large measure the result of collective work with later additions of individual qualitative results. The concepts that are implicitly present in collective thought can be revealed and made explicit in the course of teaching, in their turn accelerating not only the process of learning but also the process of research.

This idea lies behind the attention that Lawvere devotes to his students who are grappling with subjects in advanced mathematics such as algebra, analysis and geometry. Again the central theme is the *unity of mathematics*: tools useful for understanding can make clear what it is that the subjects have in common, understanding how to extract the universal properties, and knowing how to make a rigorous formal study evolve from these.

One of the characteristic traits of Lawvere's explanations is a taste for examples that are simple yet not banal, in which the fundamental ideas are put into action without the noisy interference of what is non-essential; this is emblematic of his refusal to buy into the tacit assumption that certain topics are "too advanced" or "too complicated" to teach.

Two books capture pretty completely this attitude of Lawvere's regarding the freedom and sense of confidence in ultimate success that are acquired when fundamental concepts are made explicit, no matter how complicated or advanced they might appear to be.

Conceptual Mathematics: A first introduction to categories (Cambridge University Press, 1997), co-authored with Lawvere's friend and colleague Steve Schanuel, is the result of a series of courses taught to first-year students at the University of Buffalo. The book assumes only a knowledge of high-school algebra and presents numerous basic examples – for example, of *directed graphs* or *dynamic systems* – for which are developed by means of both general considerations and applications.

The aim is twofold: on one hand, to introduce fundamental notions to non-mathematicians or to students facing the study of advanced topics for the first time; on the other, to provide the basic elements of the theory of categories to researchers in various fields, such as theoretical computer science, linguistics, logic, physics, and so forth.

Sets for Mathematics, written in collaboration with Bob Rosebrugh, was published in 2003 by Cambridge University Press. The idea is that set theory, seen as algebra of functions, should be introduced and applied early as a unitary basis for the study of advanced topics: starting with an intuitive description of the usual phenomena in physics and mathematics, what is arrived at is a precise specification of the nature of the category of relevant sets. Formal study proceeds from general axioms relative to the universal properties, without however neglecting the distinctive aspects of classic sets – the "constant" sets introduced by Cantor – through the unfolding of a method that often sheds light on Lawvere's explanation. The variable sets used in geometry and analysis are therefore given the categorical models appropriate to them.

Perhaps it is the capacity to produce theories, open new roads and unify – capacities that are rare and precious in mathematicians of any epoch – joined to the willingness to develop and present the simplest – and often for this reason the most fundamental – examples, and to give serious consideration to even the facts that are apparently the most elementary, that make Lawvere a very important thinker in today's mathematics. Perhaps it is his willingness to discuss and explain, the circulation of his ideas that often takes place in ways that are direct and personal, the friendship and geniality that he spreads along with mathematics, that make him the fascinating character that he is.

His prophecies as well have to be taken seriously:

> It is my belief that in the next decade and in the next century the technical advances forged by category theorists will be of value to dialectical philosophy, lending precise form with disputable mathematical models to ancient philosophical distinctions such as general vs. particular, objective vs. subjective, being vs. becoming, space vs. quantity, equality vs. difference, quantitative vs. qualitative etc. In turn the explicit attention by mathematicians to such philosophical questions is necessary to achieve the goal of making mathematics (and hence other sciences) more widely learnable and useable.[8]

Category Theory

The notions of category, functors and natural transformations appear for the first time in 1945 in an article by Samuel Eilenberg (1913–1998) and Saunders Mac Lane (1909–2005) entitled "General Theory of Natural Equivalences". As the title

---

[8]F. William Lawvere, "Categories of Space and Quantity", pp. 14–30 in: *Structures in Mathematical Theories: Reports of the San Sebastian International Symposium, September, 25–29, 1990*, A. Díez, J. Echeverría, A. Ibarra (eds.), De Gruyter, 1992, p. 16.

of this work indicates, the authors' attention is mainly focussed on formalising the notion of "natural transformation", while the notion of "functor", a term adopted from Carnap's logic, served to explicate the general processes in which transformations take place. In its turn, the notion of "category", which redefines in mathematical terms the categories of the philosophy of Aristotle, Kant and Charles S. Pierce, served as a support for functors. It was, then, a description of existing mathematical phenomena, expressed at the required level of generality.

Thus it happened that for a couple of decades after 1945, the notion of category appeared as a language that was particular useful for describing many results in mathematics through the use of diagrams with arrows, particularly in algebraic topology (in the wake of a famous book of 1952 by Eilenberg and Henri Cartan regarding the fundamental questions of this subject) and homological algebra (in the wake of another famous book, this time by Eilenberg and Steenrod, of 1956). Further, as a consequence of the fact that many mathematicians began to use the language of categories systematically, the conviction began to gain ground that category theory had to be seen as a "third level of abstraction, the first two levels being that of quantities and that of structures, and thus it came to embody a special kind of structuralism of mathematical objects".

But category theory is not only a language useful for describing various phenomena, or a special mathematical structure. The crucial step forward was taken in 1957 when, in an article entitled "Sur quelques points d'algèbre homologique", Alexander Grothendieck incorporated the fundamental and formal aspects of homological algebra in a special type of category – an Abelian category – showing how it was possible to carry out the principal constructions and demonstrating the corresponding results in this general set up. And, in consequence, particular categories of structures, the categories of sheaves over a generalised topological space, can take the place of Abelian categories, in order to show that, for example, the methods of homological algebra can be applied directly to algebraic geometry.

Other developments led to the consideration that the systematic presence of categories in "mathematical practise" (according to the terminology used in a fundamental text by Mac Lane) is due to the notion of "adjunct functors", originally expressed in terms of categories by Daniel Kan in 1956 (and published in 1958). Thus the subject became increasingly an independent field of research and grew rapidly, also in terms of applications. In addition to its original contexts, relative to algebraic topology and homological algebra (following Lawvere's thesis of 1964) and to logic (again thanks to contributions by Lawvere and to Joachim Lambek's use of methods of category theory in proof theory).

This phase culminated, at the end of the 1960s, in the notion of "topos" on the part of Grothendieck and his school. A topos is a category of sheaves of sets on a generalised topological space: the later elementary characterisation on the part of Lawvere and Myles Tierney (1972) led to "elementary topoi" and shed light on the fact that the logical structure of these categories is so rich that it is possible to develop a large part of mathematics in them, that is, it is possible to define numerous structures internally, to carry out the necessary constructions, and to prove the majority of their properties through the use of the internal logic. An "elementary topos" can be considered a "categorical theory of sets", thus acquiring immediately a foundational value.

After the 1980s, the theory underwent further developments and applications, in relation to, for example, the birth of new systems of logic and to the semantics of programming in theoretical computer science, or the use of "higher dimension" categories (bicategories, tricategories, etc.) in the study of the so-called "quantum groups" in theoretical physics. This is proof of the fact that category theory not only makes it possible to conceptualise traditional fields in new ways, often going beyond borders that had been established for a long time, but, constantly attentive to the axiomatic method and algebraic kinds of structures, it also contributes to the coherence, strengthening and stability of particular disciplines, revealing their universal aspects and the overall conceptual context in which they are found.

Andrew Wiles

Claudio Bartocci

Andrew Wiles, born in Cambridge in 1953, has been interested in number theory since he was a child, and in particular, in Fermat's Last Theorem.

Pierre de Fermat – born in Toulouse, a lawyer by profession, and a great "mathematical dilettante" – making brief notes in the margins of his copy of *Arithmetica* by Diophantus,[1] notes that it is impossible to "decompose a cube into two cubes, or a biquadrate into two biquadrates, nor in general to divide any power above a square in two other powers of the same degree". Expressed as a formula, this means that the equation

$$x^n + y^n = z^n$$

has no integer nontrivial solutions if the exponent n is greater than 2. Fermat wrote that he had also *"discovered a truly remarkable proof which this margin is too small to contain"*.z

In all likelyhood, Fermat – who had provided a proof for the case where $n = 4$ – fell into an error that derived from the deceptive application of the method of infinite descent. The case where $n = 3$ was solved by Euler in 1753; that where $n = 5$ by Dirichlet and Legendre in 1825; and that where $n = 7$ by Lamé in 1837. Kummer proved Fermat's theorem for all regular primes.[2]

---

[1] *Observations sur Diophante*, published posthumously in 1670 (full text available on the web at http://fr.wikisource.org/wiki/Œuvres_de_Fermat). For a life of Fermat, see M.S. Mahoney, *The mathematical Career of Pierre de Fermat, 1601-1665*, Princeton University Press, Princeton, 1994 and G. Giorello and C. Sinigaglia, *Fermat. I sogni di un magistrato all'origine della matematica moderna*, Le Scienze, Milano, 2001.

[2] See André Weil, *Number Theory. An approach through History*, Birkhäuser, Boston-Basel-Stuttgart,1983 and M. Bertolini, *L'ultimo teorema di Fermat* in *La Matematica. Volume secondo. Problemi e teoremi*, C. Bartocci and P. Odifreddi, eds., Einaudi, Torino, 2008, pp 313–334.

C. Bartocci et al. (eds.), *Mathematical Lives*,
DOI 10.1007/978-3-642-13606-1_34, © Springer-Verlag Berlin Heidelberg 2011

Andrew Wiles

After studying at Oxford, Wiles began work on his doctorate at Cambridge, studying Iwasawa theory of elliptical curves under the supervision of John Coates. Having completed his Ph.D., Wiles spent a period of time at the *Sonderforschungs-bereich Theoretisches Mathematik* in Bonn before moving to Princeton's Institute for Advanced Study, where he was named professor in 1982.

Towards the mid-1980s, Frey and Ribet proved that Fermat's last theorem is a consequence of a statement known as the Shimura–Taniyama–Weil conjecture. In 1993, after some 7 years of uninterrupted work, Wiles was able to prove this conjecture for a broad class of examples, including those necessary to prove Fermat's last theorem. He announced his result to a packed crowd of listeners at a seminar in Cambridge, concluding by simply saying, "I will stop here", in his characteristic measured and unrhetorical style. Wiles's reasoning, however, contained a small flaw: in 1995, together with Richard Taylor, he finally obtained the correct proof and brought to an end the conundrum posed by Fermat.[3]

---

[3]Cf. A. Wiles, "Modular Elliptic Curves and Fermat's Last Theorem", *Annals of Mathematics,* 141 (1995), pp. 443–551; A. Wiles and R. Taylor, "Ring-theoretic properties of Hecke Algebras", *Annals of Mathematics*, 141 (1995), pp. 552–572.

Wiles has been honoured with various prestigious international prizes and recognitions, including the Wolf Prize in 1996. Today he is the Eugene Higgins Professor of Mathematics at Princeton University.

Interview with Andrew Wiles (October 2004)

Claudio Bartocci

As tradition has it, Euclid once said that "there is no royal road to geometry". In other words, mathematics is intrinsically a difficult subject, there are no shortcuts to learning it or to doing it. Do you agree with that?

Andrew Wiles

Yes. I would add also that the two parts of learning and of creating mathematics each require their own training. Some people are more talented at one than the other but neither comes without a struggle.

CB

When you decided to become a mathematician, were you attracted to mathematics mainly because of its challenging difficulty, or by some other reasons?

AW

I was captivated by mathematics from a very young age. I do not think that I understood then how hard it was! As a child a problem that takes half an hour is hard and one that you can't do until your teacher explains it is near impossible. The realization that there are many problems which no one can do comes much later. I was aware of Fermat's last theorem as a child but I did not realize just how many unsolved problems there were in the mathematical world.

CB

Nowadays it is impossible even for the most gifted professional mathematician to embrace all of mathematical knowledge. There are so many different research areas, so many specialties, that mathematics appears to be highly fragmented. Do you think it makes still sense to consider mathematics as a whole?

AW

There are mathematicians who can master a great range of mathematics but it is hard to actively pursue a very hard problem in one area of mathematics while keeping up with the rest of the subject. I think it still makes sense to consider it as one subject so long as the common grounding we have enables us, in a reasonable period of time, to delve into any particular mathematical theory when the need or the opportunity arises. The way of thinking of a mathematician is still common to all branches of our subject.

CB

What are the main challenges of today's mathematics?

AW

As a number theorist I see my field driven by the desire to solve particular problems. In the year 2000 the Clay Mathematics Institute listed seven mathematical problems that represented some of the greatest challenges left over from the twentieth century. For me old problems such as these are the most exciting challenges. The greatest three of these seemed to me to be the Riemann hypothesis, the Poincare conjecture and the P-NP problem. Others might prefer the challenges of unifying different fields or creating new ones.

CB

In your opinion, is the main impulse to the progress of mathematics given by the solution of classical problems or by the construction of new theories?

AW

It is a case of the chicken and the egg. One uses new theories to solve particular problems and the solutions of new problems spawn new theories. The validation of a new theory is usually that it can solve a classical problem that has resisted the earlier theories. So for me the ultimate test and the greatest pleasure comes from solving the classical problems.

CB

You spent 7 years, in complete isolation, to prove Fermat's last theorem. However, "publish or perish" seems to be the rule of today's science and researchers rush to submit their papers to the web archives. Please comment on that.

AW

I think the speed of mathematics publication is still well below that of the rest of science. One still has time to struggle for years over the hardest problems. However there is a psychological price to pay for that as of course one can not give evidence of this hard struggle and one may end with little to show for it. On the other hand always tackling the more reasonable problems usually has the obvious result one only solves the reasonable problems. Each mathematician has to choose a mode of working that they can live with.

CB

In a letter to Robert Hooke, Isaac Newton wrote, "If I have seen further, it is by standing on the shoulders of giants". What's your personal experience?

AW

I am aware that for the first 300 years no one could have solved the Fermat problem by the method I used. It is simply built on too much modern mathematics. Even 20 years before I did it the problem would have been much much harder. There is a great deal of luck involved in living at the right time! And the problem is that one does not know whether one is living at the right time. Is it possible for someone now to prove the Riemann hypothesis? I believe so but I don't know for sure. So the answer is that certainly the solution of Fermat depended on the work of many many others, including perhaps many who were not giants.

CB

André Weil made the following remark in his short essay *De la métaphysique aux mathématiques:* "*Rien n'est plus fecond, tous les mathematiciens le savent, que ces obscures analogies, ces troubles reflets d'une théorie à une autre, ces furtives caresses, ces brouilleries inexplicables rien aussi ne donne plus de plaisir au chercheur*" ("Nothing is more fruitful, as all mathematicians know, than those vague analogies, those obscure reflections that lead from one theory to another, those furtive caresses, those inexplicable traces: nothing gives more pleasure to the researcher"). Do you agree that analogy has a key role in mathematical discovery?

AW

Yes I do and especially in number theory. There is so little natural geometric intuition to use, so little of the real world, that one is forced to conjure up the most tenuous analogies. Sometimes when you try to explain them to another mathematician they almost evaporate.

CB

In most European and American universities the number of math (and more generally, science) students is constantly decreasing. What would you say to a young person to convince him or her to study mathematics?

AW

I believe that to lead a satisfying life you have to pursue something that you are passionate about. It is not enough to be good at mathematics though it certainly helps. You have to really love doing it. You have to feel an urge, for example when waiting for a train to move, to pick up a piece of paper and start working on your latest problem. Only such a passion can keep you going when you get the inevitable frustrations of being stuck in a difficult part of the problem. As a mathematician you will be part of a community that has existed for thousands of years and you will contribute to a creative enterprise that spans the centuries and civilizations. But life is too short to be wasted pursuing something you do not care about. So only do it if you love it.

Mathematical Prizes

The Fields Medal and the Abel Prize

There are prizes that are awarded periodically to honour the best mathematicians, both the most promising and those who have already enjoyed a remarkable career. Here we note the two most important of such prizes.

The Fields Medal

The Fields Medal is a prize given every 4 years by the International Mathematical Union on the occasion of its International Congress. The prize was founded in 1936 and has been awarded regularly since 1950. The prize is named for John Charles Fields, secretary of the 1924 International Congress of Mathematicians in Toronto, who provided the funding necessary to mint the special gold medals. The prize is intended to recognition younger mathematical researchers who have made significant contributions, and at the same time provide financial support for future work. For this reason, the medal is conferred on mathematicians who are not yet 40-years old at the time of the Congress. In consideration of the rapid expansion of mathematical research, in 1966 it was decided to present the medal to up to four mathematicians. Here is a list of the winners:

1936 (Oslo): Lars Ahlfors, Jesse Douglas
1950 (Cambridge, USA): Laurent Schwartz, Atle Selberg
1954 (Amsterdam): Kunihiko Kodaira, Jean-Pierre Serre
1958 (Edinburgh): Klaus Roth, René Thom
1962 (Stockholm): Lars Hörmander, John Milnor
1966 (Moscow): Michael Atiyah, Paul Joseph Cohen, Alexander Grothendieck, Stephen Smale
1970 (Nice): Alan Baker, Heisuke Hironaka, Sergei Novikov, John G. Thompson
1974 (Vancouver): Enrico Bombieri, David Mumford
1978 (Helsinki): Pierre Deligne, Charles Fefferman, Grigory Margulis, Daniel Quillen

C. Bartocci et al. (eds.), *Mathematical Lives*,
DOI 10.1007/978-3-642-13606-1_35, © Springer-Verlag Berlin Heidelberg 2011

1982 (Warsaw): Alain Connes, William Thurston, Shing-Tung Yau

1986 (Berkeley): Simon Donaldson, Gerd Faltings, Michael Freedman

1990 (Kyoto): Vladimir Drinfel'd, Vaughan F. R. Jones, Shigefumi Mori, Edward Witten

1994 (Zürich): Jean Bourgain, Pierre-Louis Lions, Jean-Christophe Yoccoz, Efim Zelmanov

1998 (Berlin): Richard Borcherds, Timothy Gowers, Maxim Kontsevich, Curtis T. McMullen

2002 (Beijing): Laurent Lafforgue, Vladimir Voevodsky

2006 (Madrid): Andrei Okounkov, Terence Tao, Wendelin Werner

2010 (Hyderabad): Elon Lindenstrauss, Cédric Villani, Ngô Bao Châu, Stanislav Smirmov

(In 1998, Andrew Wiles was presented a silver plaque in special recognition; in 2006 Grigorij Perelman declined the prize).

The Abel Prize

The Abel Prize is an international award presented annually as of 2003 by the Norwegian Academy of Sciences in recognition of excellent scientific work in the field of mathematics. The prize amounts to six million Norwegian crowns (about €750,000), and is funded by the Niels Henrik Abel Memorial Fund, an endowment specially created by the Norwegian government to encourage scientific research and education. According to the by-laws, in addition to funding the prize, the endowment is to be used to finance scientific activities aimed at young people.

Here is a list of the Abel Prize Laureates up to 2009, along with the jury's comments:

2003 – Jean-Pierre Serre: "for playing a key role in shaping the modern form of many parts of mathematics, including topology, algebraic geometry and number theory"

2004 – Michael F. Atiyah and Isadore Singer: "for their discovery and proof of the index theorem, bringing together topology, geometry and analysis, and their outstanding role in building new bridges between mathematics and theoretical physics"

2005 – Peter Lax: "for his groundbreaking contributions to the theory and application of partial differential equations and to the computation of their solutions"

2006 – Lennart Carleson: "for his profound and seminal contributions to harmonic analysis and the theory of smooth dynamical systems"

2007 – S. R. Srinivasa Varadhan: "for his fundamental contributions to probability theory and in particular for creating a unified theory of large deviation"

2008 – John G. Thompson and Jacques Tits: "for their profound achievements in algebra and in particular for shaping modern group theory"

2009 – Mikhail Gromov: "for his revolutionary contributions to geometry"

2010 – John Tate: "for his vast and lasting impact on the theory of numbers"